CONTENTS

LIST OF FIGURES AND TABLES

INTRODUCTION

In recent years, artificial intelligence has moved from the realm of science fiction into a force reshaping our everyday lives. The rapid evolution of AI has sparked both awe and debate, often portrayed in the media as a double-edged sword with immense potential and profound implications. But what if we learn to harness this technology to our advantage instead of fearing the rise of intelligent machines? Imagine a world where AI does not confine us to a matrix of monotonous tasks but liberates us to explore our highest human potential—creativity, strategic thinking, and meaningful connections. This book is your guide to understanding and mastering AI, not just to navigate the digital age but to thrive in it. Here, you will discover how to strategically use AI to break free from the constraints of routine and unlock new opportunities in both your personal and professional life.

Imagine you are at home, surfing the internet, watching a movie suggested by an algorithm, and checking your health metrics on an app. Artificial intelligence (AI) is seamlessly interwoven into these applications, operating right under your nose. Today, AI's influence spans many diverse activities, from transforming media consumption to revolutionizing healthcare and finance. It is not just a trend but an integral part of our everyday lives, rebooting how people interact with the world, often at an unconscious level. AI curates and recommends your favorite entertainment, manages your investments, and monitors your wellness, adjusting to your tastes and needs with incredible accuracy.

Imagine a future where AI is as unnoticeable and essential as electricity. This is not fiction but a real possibility in the next decade, highlighting how AI is becoming ingrained in every aspect of our lives. Tomorrow promises instruments that will remove language barriers and introduce innovations once reserved for fairy tales. Real-time language translation will enable seamless communication across different languages. Intelligent personal assistants, fully autonomous vehicles, highly realistic virtual realities, and other futuristic technologies will enhance everyday life in ways that seemed fantastical before. AI-automated activities will become embedded in the fabric of our daily lives, making technology virtually invisible, intuitive, predictive, and perfectly integrated.

While these advances are remarkable, they also bring unique challenges and questions. Where do we go from here? What is the impact on privacy, employment, and creativity? You may find yourself at a crossroads—you are curious, perhaps also a little apprehensive, seeking answers about this technology that can fundamentally shape your future. You may wonder how AI makes decisions, who is liable when it fails, and what the workplace will look like in an increasingly automated world. These are valid concerns as we move toward a future where AI will become an even more indispensable part of our daily lives and work.

This book will delve into the technical innovations that AI is introducing and their societal impacts. It will explore the comprehensive changes this technology will bring, empowering you to make informed decisions and capitalize on emerging opportunities. It will guide you through the intricacies of generative AI, the secrets of large language models (LLMs) and ChatGPT, and the fundamentals of deep learning. By examining the interplay of AI with various sectors, including machine learning and the emerging field of quantum computing, you will gain valuable insights into the principles driving AI's advancements. Understanding machine learning is crucial because it forms the backbone of AI, enabling systems to learn from and adapt to data. Quantum computing, on the other hand, promises to exponentially enhance AI's capabilities by solving complex problems faster than classical computers ever could.

You will gain insights into the ethical considerations for the responsible use of AI, enabling you to make informed decisions in today's fast-changing digital age. Topics include data privacy, algorithmic bias, and the moral implications of automated decision-making.

Whether you are an executive using AI to gain an edge against the competition, a student keeping up with rapid innovations, or an enthusiast curious about this technology, this book offers critical insights to enhance your understanding and decision-making. It aims to transform how you use AI to improve the quality of your life and work and gain a strategic advantage.

AI entails a mix of algorithms and machine learning capabilities designed to perform tasks that typically require human intelligence. From Siri's speech recognition to Tesla's autonomous vehicles, AI is achieving feats that were unimaginable just ten years ago. This book will demystify the jargon and bring you face-to-face with the applications and innovations that have redefined how we interact and work, showing how AI has become integral to everyday technology.

It will provide you with a solid understanding of artificial intelligence's capabilities and position you to maximize its potential in both personal and professional spheres. Visualize yourself mastering the technologies shaping the next wave of innovation, stepping into a future where AI supports and sometimes surpasses human ability. You will learn to apply this technology efficiently, discovering new opportunities and creating unprecedented solutions.

In the past, understanding AI was reserved for experts, but no longer. This book will be your trusted companion, breaking down barriers and revealing the steps to master its practical use. It offers a straight-forward, non-technical approach to understanding and applying AI while enabling you to assess its impact on society and the economy. Whether you aim to advance your career, deepen your skills, or better understand this technology, this book will equip you with the tools and knowledge to achieve your goals.

Beyond merely informing you about artificial intelligence, this book will also lay a foundation for expanding your engagement with the exciting possibilities AI offers. It stands out by explaining this most pivotal technology of our lifetime in a familiar, understandable, and accessible way, making it an essential tool for navigating our digital age.

1

THE MAGIC BEHIND GENERATIVE AI

The Nature of Generative AI

Generative AI represents the front line of the vast artificial intelligence technology realm. It is programmed to generate new material from given datasets. It can create text, images, audio, complicated designs, and other content (McKinsey & Company, 2024b). If you ask an AI model such as ChatGPT a question, it will respond with a contextually appropriate and relevant text based on what it has learned during its training. This example not only shows the practicability of generative AI but also points to its transformative capability in different sectors of the economy. Before we jump into the fascinating range of generative AI applications and their impact, it is essential to understand how the technology works (Lawton, n.d.).

The Science of Generative AI

Machine learning, which underpins generative AI, is a fundamental science that teaches computers to learn from and make data-based decisions. This chapter will investigate these systems' learning processes and the intricate mechanisms used to develop content that resembles human creativity and insights.

Learning from Data

Generative AI works by using models trained on large and complex datasets. These models, usually deep learning networks that resemble human neural pathways, train on big datasets for different purposes, including written texts, images, sounds, and videos. The first step in AI learning is feeding data into the model. During this training period, the model learns to identify and analyze the patterns and structures contained in the data (Conn, 2023).

Pattern Recognition and Statistical Learning

Generative AI can create novel content thanks to its ability to recognize patterns. By using advanced statistical methods, these models can reveal the latent structures in the data they are processing (Zhou & Lee, 2024). This ability is imperative because it is the foundation of all subsequent content-creation activities. Take, for instance, a generative model trained on musical scores. It learns not only the notes and rhythms but also the styles and subtleties that differ among the different music genres and artists, after which it can emulate them (*The Generative AI Technology Stack,* n.d.).

Algorithms and Neural Networks

The algorithms that drive generative AI work by using neural network architectures, including the following:

- **Convolutional Neural Networks (CNNs)** are incredibly efficient at processing pixel data and are widely used to visualize or recognize images.

- **Recurrent Neural Networks (RNNs)**, and notably Long Short-Term Memory (LSTM) networks, can handle sequences, like language for text generation or time series data for projection and predictions.

- **Transformer Networks** are effective in parallel processing and able to handle long-range dependencies in data. They are instrumental for neural networks in models such as GPT (Generative Pre-trained Transformer) that produce human-sounding language.

Generative Models

Three primary types of generative models play significant roles in AI.

- **Generative Adversarial Networks (GANs)** contain two neural networks, a generator and a discriminator, which work antagonistically against each other. The generator creates data in images while the discriminator evaluates their authenticity. The perpetual interaction between these two components enhances an AI model's power to produce increasingly high-quality output, such as lifelike and photorealistic images.

- **Variational Autoencoders (VAEs)** reduce data into a compact representation and then restore new data instances from the representation. VAEs are helpful in cases where data preservation is crucial, such as medical imaging analysis.

- **Autoregressive Models** generate data one element at a time, predicting the next element based on previous ones. Transformers like GPT are a prominent example.

We will return to these main types of generative models later to explain in more detail how they work.

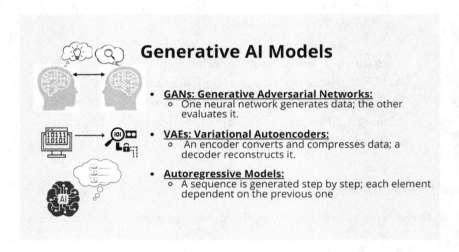

Fig.1: The Three Main Kinds of Generative AI Models.

Iterative Improvement and Learning Efficiency

Generative AI models improve through iterative processes; each training cycle refines the outputs based on the accuracy and quality of the feedback provided. This iterative process of minimizing errors and gradually increasing the model's accuracy is called backpropagation. Besides the transfer learning method, which adapts a model trained for one task to another, and the reinforcement learning method, which teaches models to make sequences of decisions, other methods are also used to train generative AI systems.

Key Technologies in Generative AI

Generative AI is the result of the advancement of certain breakthrough technologies that allow it to accomplish high-level tasks ranging from photorealistic image creation to coherent and contextually appropriate text production (Takyar, n.d.). Reviewing these technologies will give us an understanding of how generative AI models are built and achieve outputs that habitually rival those created by humans (Zderic, 2023).

Neural Networks

Generative AI revolves around neural networks based on the biological structure of the human brain. These networks comprise layers of interconnected nodes (neurons), where each node processes inputs and passes on its output to the following layers. The complexity and depth of these layers enable neural networks to model high-level abstractions and intricate patterns in data. This feature is essential for tasks involving an understanding of visual perception and natural language processing.

Generative Adversarial Networks (GANs)

GANs can be considered to be a new and potent type of neural network for generative AI. Introduced by Ian Goodfellow in 2014, GANs consist of a generator that creates images or other data types from a random noise input and a discriminator that differentiates generated images from real ones. The generator and discriminator train competitively, and the generator tries to create more accurate and realistic outputs to deceive the discriminator. This approach allows GANs to achieve high-quality results, making them a potent tool for generating realistic images, enhancing photos, and creating art. We can compare the activity of GANs to the thought processes that happen when two people are engaged in a critical dialogue that hones and refines ideas as the output of their conversation.

Variational Autoencoders (VAEs)

VAEs are another type of neural network designed for generative tasks. Unlike GANs, they code data into a compressed representation, from which an encoder and decoder recreate content. The encoder maps the input data into a smaller, sparser representation, and the decoder reconstructs the input from its compressed form. During the training cycle, the difference between the original input and its reconstruction is minimized, so the decoder generates acceptable and readable outputs. Simply put, we can make an analogy between VAEs and the way human memory works. It also encodes experiences and information in a more abstract, compressed form. Instead of remembering every detail, our brains store essential features and patterns to apply later.

Transformer Networks

Transformers are the latest model category that fundamentally changed how we approach tasks such as natural language processing (NLP). In their paper, Attention is All You Need, Vaswani et al. (2017) posit the transformer model, which is based heavily on attention mechanisms that weigh the impact of different parts of the input data. Contrary to previous models that processed data sequentially, transformers use parallel multisets for faster training and better management of long-range dependencies in the data. Like the human brain focusing on key aspects of a conversation simultaneously to understand the context and meaning, transformers use attention mechanisms to weigh the importance of different parts of the input data.

Reinforcement Learning

Reinforcement learning (RL) is a form of machine learning where an agent learns by performing actions and receiving rewards or penalties in return. In the context of generative AI, RL can enhance models by continuously refining their actions (e.g., adjustments to generated content), ultimately leading to higher rewards. Rewards refer to measurable benefits, such as increased user engagement or immersion. This approach is particularly effective when output quality improves through trial and error, such as video game design or robot movements. Continuing the human analogy, we can compare RL to behavioral conditioning, which involves learning through rewards and punishments based on actions taken.

Deep Learning Optimization Techniques

Training uses various optimization techniques to increase the performance and efficiency of AI models.

- **Dropout** prevents overfitting by randomly deactivating neurons during training. It is like periodically focusing on different aspects of a problem to avoid overly relying on a single perspective.

- **Batch normalization** normalizes inputs to each layer, improving stability and speed. It is akin to maintaining a balanced and consistent mindset while learning new information.

- **Adam (Adaptive Moment Estimation) optimizers** are a powerful optimization technique in machine learning that adaptively adjusts learning rates for each parameter based on moment estimates. This approach enhances the speed and stability of convergence—the process where the model's learning efficiency stabilizes and reaches optimal values—

making Adam optimizers particularly useful for training complex generative AI models.

Generative AI vs. Other Forms of AI

Mastering the distinction between generative AI and other types of artificial intelligence is crucial to understanding the depth and possibilities of AI technologies. The primary distinction lies in their core functionalities: Generative AI is all about creation, whereas traditional AI usually deals with analysis and interpretation (Wang, 2024). It can be helpful to compare the two from various standpoints.

The Nature of Output

Generative AI develops new data instances similar to the original datasets. For example, generative AI may produce images that look exactly like photographs of real objects or scenery, write stories, compose music, or create realistic human speech. The outputs go beyond generating new combinations of existing data and produce new creations that did not exist before (Sweenor, 2023).

Traditional AI is also called "discriminative AI." These systems use input data for analysis and decision-making. For instance, traditional AI can recognize objects within an image, classify emails as spam or not, or predict customer behavior based on past purchases. In this case, the output usually consists of a classification, decision, or prediction based on the analysis of input data.

Data Handling and Processing

Generative AI uses complicated models that can recognize and reproduce data distributions in a data set. It identifies and selects key data components, such as structure, style, and underlying patterns, and then utilizes this information to create similar outputs (Ayuya, 2024).

Traditional AI emphasizes pattern recognition and decision-making based on predefined rules or learning patterns derived from data without creating new content. Instead, trained models are used to interpret and classify existing information.

Applications

Generative AI has been fundamental in the creative and design fields, in the genesis of art, music, literature, and even virtual environments. It also plays a crucial role in data augmentation, creating new datasets to train other machine-learning models and improve their accuracy and robustness (Shulman, 2024).

Traditional AI typically powers applications that require fast and correct decisions, including autonomous driving systems, financial forecasting, fraud detection, and medical diagnostics. Such applications use AI's capacity to process vast amounts of data to generate accurate and practical recommendations.

Methodological Approach

Generative AI usually starts with models such as GANs (Generative Adversarial Networks) or VAEs (Variational Autoencoders) built with learnable data distributions in mind.

Traditional AI utilizes models such as decision trees, support vector machines (SVM), and linear regression, which categorize data into previously defined classes and predict precise outcomes.

Decision trees split data into branches based on feature values are like making a series of yes/no decisions to navigate a flowchart. Support vector machines find the optimal boundary to separate different classes by maximizing the margin between them, improving the model's ability to generalize from the training data to new, unseen data crucial for predictive performance. Linear regression, which predicts outcomes based on linear relationships between variables, is similar to drawing a straight line through data points to estimate trends.

Innovation vs. Optimization

Generative AI stretches the limits of what machines can produce by creating something new, thereby driving innovation. It pushes the boundaries of computation and provokes our creative imagination, which belongs to the realm of art and humanity.

Traditional AI focuses on process optimization and decision-making improvement in multiple sectors. It does not alter existing systems but enhances their efficiency and accuracy within established frameworks and conditions.

The Applications of Generative AI

Generative AI is reshaping many sectors through its ability to generate and create. From redefining the domains of art and design to changing the channels of content creation and driving new product development, this technology is proving to be a game changer across many industries.

AI in Art and Design

Generative AI has quickly carved a niche in the artistic and design industries, enabling professionals from these sectors to stretch the limits of their imagination. Through algorithms that work on vast numbers of artworks, generative AI can create art that reflects the same emotional and aesthetic elements as that made by humans. For instance, AI tools can produce complicated digital paintings, distinctive furniture, or even fashion lines. These abilities give impetus to artistic creativity and enable designers to explore complex patterns and structures that would otherwise be hard to visualize without first risking the trial and error of full production (Borges, 2023).

Real-World Applications

The introduction of generative AI within art and design has inspired and led to the development of practical applications reinventing these professional fields. Several notable real-world examples highlight AI's diverse capabilities in these fields.

- **Automated Graphic Design Tools:** Some companies, such as Canva and Adobe, have introduced AI into their platforms in order to automate and improve the process of design tasks. For instance, Canva's "Magic Resize" tool uses AI-driven technology to dynamically resize designs to fit various formats and dimensions, thus making it much easier for users to create content across a wide range of platforms. Sensei AI by Adobe enables professional photographers to edit complex tasks like object selection and apply them rapidly.

- **Customized Furniture Design:** Companies like Cazza and SketchChair are harnessing the power of AI to transform the furniture industry. These platforms allow users to input their preferences and constraints. AI algorithms then suggest design options that are both aesthetically pleasing and functional. This approach enables mass furniture production customized to individual preferences, making personalized designs accessible to a broader audience.

- **Fashion Design:** AI is essential for trend prediction and design creation in the fashion industry. Tools like Stitch Fix's algorithm analyze customer data to forecast future trends and recommend individualized apparel choices. In addition, designers use AI to generate inspiring patterns and fabrics that can create great new fashion pieces with a unique look.

- **Architectural and Urban Planning:** AI algorithms are applied in the design of buildings and the arrangement of urban areas. Companies like Spacemaker use AI to provide tools for architects to build as spatially as possible while at the same time observing zoning laws and environmental constraints and maintaining aesthetics. This technology is able to generate thousands of configurations in a short period to discover the most economical and sustainable designs.

- **Digital Art Creation:** Artists can use platforms like RunwayML to incorporate AI into their creative processes, producing complex and fascinating works of art that would otherwise be impossible to create by hand. AI has already established itself in the art market, selling pieces at major auction houses like Christie's, indicating the burgeoning demand for algorithmically generated art.

- **Interactive Media and Entertainment:** In the entertainment industry, AI creates adaptive media experiences that respond to user inputs. Video games and virtual environments apply AI to develop dynamic content that adjusts to players' actions, resulting in a more exciting and personalized user experience.

How AI is Changing the Content Landscape

AI powered by generative technology is becoming a game changer in content creation. It writes articles, composes music, and even script videos, thus changing the media production landscape. Media companies utilize AI to produce financial reports and sports recaps, allowing human journalists to cover more in-depth stories. Likewise, artificial intelligence tools aid musicians by generating new melodies that serve as a fresh source of inspiration for their creative work.

Real-World Applications

Recently, AI technologies like OpenAI's ChatGPT and Google's AI-powered writing assistants have been recasting the landscape of content creation. These tools exemplify how AI is leveraged to streamline production processes, enhance creativity, and introduce new forms of storytelling (Fortino, 2023).

- **Automating Routine Writing Tasks:** AI-driven writing aids are finding more applications to handle repetitive writing tasks like reports, summaries, and articles based on data, freeing journalists to do more extensive reporting and analysis, which expedites content generation and ensures high precision and consistency.

- **Enhancing Editorial Processes:** Writers and editors integrate AI tools like Grammarly and Hemingway Editor into their workflow to improve content quality. These resources provide instant grammar, style, and readability recommendations to allow corrections and compliance with editorial standards, which produces professional-grade writing that more effectively grabs readers' attention (Eric, 2024).

- **Shaping Interactive and Personalized Content:** In the future, AI will lead to personalized content development. Platforms like Heliograf, developed by The Washington Post, are being employed to produce personalized stories and reports that meet readers' interests and preferences. Another example is responsive AI from the gaming and virtual reality space. Its adaptive narrative can change based on a user's decisions, resulting in a highly personalized and interactive experience.

- **Supporting Creative Writing:** In the creative domains, AI tools like ChatGPT generate ideas, write stories, and even pen poems and help write novels. These resources can propose plot twists, dialogues, and character developments, which is particularly helpful for breaking up stagnated thinking patterns or transforming initial ideas into a vivid, complex story.

- **Scriptwriting and Video Production:** The film and video industry is applying AI to generate scripts and help in the pre-production process. AI can evaluate scripts to anticipate how an audience would react, recommend modifications, and even help directors visualize scenes before the shooting starts. These tools can cut production time and cost while enhancing storytelling (Klok, 2024).

- **Educational Content and E-Learning:** AI is disrupting educational content by driving the creation of bespoke educational experiences. Platforms that intelligently modify content can be applied to alter the style and pace of learning, making it easier and education more effective. AI also creates materials like interactive tutorials and quizzes that can be adjusted and fed back to students based on their individual understanding and abilities.

Generative AI in Innovation and Product Development

Generative AI is at the core of innovation and product development, especially in the pharmaceutical, automotive, and consumer electronics industries. By generating and analyzing different designs of a product's or drug's molecular structure, AI can significantly increase the speed of the R&D process, cutting down on time to market and increasing the input of new ideas that human designers could not otherwise conceive.

Real-World Applications

Another important application is to predict and evaluate drug compound effectiveness, which in the past could only be achieved with much effort and funding.

- **Prediction of Drug Efficacy:** AI algorithms can analyze all available libraries of molecular structures and their pharmacological features to estimate the efficiency of new drug molecules. Through the study of past data, these models will be able to predict how particular compounds will respond to the presence of specific biological targets, which may lead to the discovery of medicine that is more effective than what can be developed by traditional methods (McKinsey & Company, 2024a).

- **Reduction of Laboratory Testing:** Generally, designing a new drug involves synthesizing many molecules in a lab and evaluating their safety and effectiveness. AI accomplishes this by screening various possibilities before they reach the laboratory level, accelerating drug development, and decreasing costs and the use of resources.

- **Enhanced Precision Medicine:** AI techniques are also critical for developing precision medicine, focusing on treatment based on an individual geneset. AI can use patients' genetic data to determine the most suitable drugs, considering their distinct biological markers. This approach leads to more precise and individualized treatments that are more likely to be effective, thereby enhancing treatment results.

- **Simulations and Virtual Screening:** Besides predictive analytics, AI enables virtual drug trials and simulations that predict how drugs may behave in a virtual human body model. This allows researchers to use computer models to understand drug behavior better and speed up the preclinical stages of drug development without first having to do animal tests.

- **Integration with Other Technologies:** AI and other technologies, such as robotics and high-throughput screening, frequently interact in the pharmaceutical industry. For instance, AI can navigate robotic systems to run automated experiments on specific compounds, processing data at speeds and volumes that a human scientist cannot achieve (Fraenkel & Kamath, n.d.).

- **Real-life Case Study "Atomwise":** The AI-based drug discovery company Atomwise uses deep learning algorithms to predict how different chemical compounds will affect particular targets. The AtomNet technology, which discovers drugs for Ebola and multiple sclerosis, is faster than conventional approaches (Adl, 2023).

The Future of Generative AI: Interactivity

Generative AI holds great promise for the future. As the technology improves, more cutting-edge applications will emerge that offer personalized and adaptive learning environments in education, enhanced virtual reality environments, and more efficient urban planning. As AI continues to improve in interpreting human emotions, we could witness its deeper integration into our personal interactions, introducing new methods of communication and self-expression.

Expanding Real-World Applications

The potential of generative AI goes beyond its current applications. Notably, adding more interactive personal experiences has the potential to impact the entertainment sector dramatically. An exciting development is the emergence of interactive movies in which the storyline adapts to a viewer's reactions and choices in real time, offering everyone an innovative narrative.

- **Personalization of Plot and Characters:** Generative AI can customize character maturation, plot twists, and dialogue by analyzing a viewer's past interactions, preferences, and emotional responses, promising to make experiences more engaging and personal. For example, suppose the AI picks up on a user's preference for mystery rather than romance. In that case, it will use dynamic storytelling to include more elements of suspense while still keeping the plot consistent (McKinsey & Company, 2023).

- **Real-time Reaction Analysis:** Progress in emotion recognition technologies and biometric sensors will allow generative AI to analyze real-time data from viewers, like facial expressions, heart rate, and even vocalizations, to affect the plot dynamically. This interactivity would transform storytelling into a dialogue between the viewer and the narrative, where the viewer's emotional and physical response dictates the story's course.

- **Enhanced Immersive Experiences:** Generative AI technology combines VR or AR to create immersive and interactive movies. This way, the plot could develop according to a viewer's choices while the environment alters, too (Frey & Osborne, 2023).

- **Collaborative Storytelling:** AI can let several viewers impact the storyline in tandem as it unfolds in real time. Inputs from every participant can be integrated to create a collaborative, immersive experience that reflects the collective feedback and reaction of the group, thereby making social media interactions about content consumption more interactive (Ooi et al., 2023).

Key Challenges and Ethical Considerations of Generative AI

With the growth of generative AI and its integration into different sectors, many complex problems and ethical issues could arise. Among the most significant challenges are copyright and ownership aspects and the biases embedded in AI systems.

Copyright and Ownership Considerations

Generative AI presents a richly layered mix of challenges and opportunities to present IP policies that may impact copyrights, contributions, and the creative industries' economies. These elements are critical as AI enters new sectors, and we increasingly depend on them for many forms of content creation (Holloway et al., 2024).

INTELLECTUAL PROPERTY RIGHTS

Complexities in Ownership

Traditional IP laws are straightforward; they assume human creators of novel products. On the other hand, when an AI performs content generation, the issue of authorship becomes complicated. As a tool, AI has no legal personality and cannot own its products in most laws, which begs the question of who owns AI-generated content—is it the programmer, the user, the company that offers an AI application, or the public?

Need for New Legal Frameworks

Recognizing AI-created content requires adherence to, and sometimes adjustments of, current legislation on intellectual property (IP). Some legal scholars and public officials argue in favor of a special type of copyright adapted to AI, which could be an effective framework that recognizes AI's contribution and protects human interests simultaneously. Take, for instance, the European Union, which is exploring policies that reserve a particular status for AI-generated products to balance innovation and copyright norms (Dilmegani, 2024).

ATTRIBUTION AND CONTRIBUTION

Data Usage and Compliance

AI systems require enormous datasets during their training phase, which may include copyrighted works. It is crucial to use this data legally, acquire the proper licenses, and design AI-powered IT systems that comply with legal requirements. When AI generates content using this data, it should be clear how much of the output is derived from copyrighted material and how much is uniquely generated by the AI.

Clear Demarcation of AI and Human Contributions

In areas such as academia and journalism, where the authenticity and originality of content are of the utmost importance, we must design clear regulations that enable differentiation between human and AI contributions. The role of AI assistance should be acknowledged to ensure transparency and authenticity, which are essential in research and content creation.

ECONOMIC IMPACTS

Shifts in The Creative Industries

The ability of AI to create complex creative work quickly and at a low cost fundamentally impacts the economics of creative fields. Traditionally human-made products may lose market value as AI increasingly handles routine and complex creative tasks, affecting the remuneration and livelihoods of human creative work.

Protecting Traditional Creators

Administrators should develop policies that support artists and other creatives to resolve economic gaps. These could include compensating them for AI-generated works based on their original content or providing incentives for artists to integrate AI innovations with their human creativity.

Navigating Economic Justice

AI should be integrated into the creative industry's processes to foster economic justice and maintain cultural diversity. Policymakers should focus on AI's implications for various social groups, ensuring its benefits in the creative industries do not widen the social divide.

Bias in Generative AI

As AI advances and is integrated into many sectors, addressing the biases within intelligent systems is paramount. We will highlight several key areas where bias can manifest in generative AI and the challenges associated with each.

Data Bias

Bias in data occurs when the datasets used to train AI models include skewed data. This can appear due to historical inequalities, demographic imbalances, or subjective data collection methods. For example, if literary works that reflect historical gender or racial stereotypes are used to train AI, this may, in turn, lead to the emergence of biases. This difficulty comes from the fact that these biases are often hidden and widespread, which is usually a reflection of deeply rooted social norms that are hard to identify and remove from the data (*Addressing bias in AI*, n.d.).

Algorithmic Transparency

The opaqueness of many AI systems' decision-making processes is often referred to as a "black box" problem, which makes tracing how conclusions are reached difficult, particularly in systems with intricate algorithms like deep learning. This challenge is even more severe in departments where AI's decisions have enormous consequences for individuals, such as healthcare, criminal justice, and employment, where determining the foundation of AI's decisions is fundamental for a just and fair society.

Ethical Use and Misuse

Due to its realistic and high-end outputs, generative AI may fuel ethical issues and misuse. One striking example is the development of deep fakes—very realistic and thus easily believable fake videos or audio recordings that disseminate false information, imitate individuals, and are used to perform fraud. While there are increasing opportunities to use AI technologies safely and securely, there are also questions related to privacy, consent, and the spread of disinformation.

Inclusivity and Equity

We must consider the possibility that, if not monitored, AI systems may not be equally effective for all population segments, creating systems that unintentionally advantage certain groups to the disadvantage of others, reinforce inequality, or even worsen it. The difficulty is ensuring that AI systems work fairly and equally since the historical data that they work with may not represent diversity.

Key Takeaways

- **Omnipresence of The Technology:** Generative AI is a transformative technology that significantly impacts daily life via its integration into everyday applications, such as personalized recommendations on streaming services, AI-generated content in social media, and automated customer service chatbots. These applications demonstrate how generative AI enhances user experiences and streamlines interactions with digital media.

- **Generative AI Leverages Advanced Technologies:** Generative Adversarial Networks (GANs), Variational Autoencoders (VAEs), and reinforcement learning contribute uniquely to creating new data and content. GANs are particularly notable for their ability to generate realistic images and videos, VAEs for their role in efficient data compression and generation, and reinforcement learning for optimizing decision-making processes in dynamic environments.

- **Traditional vs Generative AI:** Unlike conventional AI, which focuses on recognizing patterns and making predictions based on existing data, generative AI creates new data and content. This distinction allows generative AI to drive innovations in creative industries, such as art, music, and entertainment, by producing novel and unique outputs that were previously unimaginable.

- **Domain Agnostic Technology:** Generative AI applications span many industries and have the potential to transform whole sectors. The technology is particularly relevant in the creative industries and R&D. For instance, generative AI assists in drug discovery and the creation of personalized treatment plans in healthcare.

- **Ensuring Responsible Use:** Despite its vast potential, generative AI presents significant ethical challenges, particularly concerning bias and transparency. Addressing these issues is crucial to ensure that AI systems are fair, accountable, and transparent in their operations. It is essential to develop robust frameworks and guidelines to mitigate bias and promote the ethical use of AI technologies.

Generative AI represents a significant leap in the capabilities of artificial intelligence, enabling innovations and creativity previously exclusive to humans across various fields. However, we must rigorously examine generative AI's ethical and bias-related challenges during the development and deployment stages. Ensuring fairness, transparency, and accountability in the use of AI is crucial as we make ever greater use of this powerful technology. Despite its vast potential, we must emphasize the responsible governance of generative AI to maximize benefits and mitigate risks. As generative AI continues to evolve, society must carefully establish clear boundaries and define the role that it will be allowed to play.

2

DEMYSTIFYING LLMS AND CHATGPT

Understanding Large Language Models

What are LLMs?

Large Language Models (LLM) represent a milestone development in artificial intelligence. These models are highly intelligent algorithms that generate and process natural language similarly to humans (Amazon Web Services, 2024).

Core Functionality

LLMs are grounded in deep learning, a branch of machine learning that uses neural networks with numerous layers trained on vast amounts of data. The "large" in LLMs refers not only to the extensive size of these neural networks, which consist of millions or even billions of parameters, but also to the large datasets used in their training. The training process involves immense quantities of text from books, articles, webpages, and other text-rich sources.

Training Process

During training, LLMs learn to predict the next word in a sequence based on the words that come before it. Unsupervised learning techniques teach models to detect patterns and relationships in the data without being explicitly told what to do. With every prediction, they become more accurate, increasing their performance in future iterations.

Generative Capabilities

The ability of LLMs to predict the next word in a sequence ensures that the generated text is cohesive and contextually aligned with the input. For instance, when prompted with "The quick brown fox," an LLM might respond with "jumps over the lazy dog," thanks to its training on similar sentence structures. This capability enables LLMs to complete sentences, paragraphs, and even entire documents in a way that reads like natural human language.

Contextual Understanding

What differentiates LLMs from previous models is their contextual and situational competence. Most modern LLMs reflect a transformer architecture that utilizes "attention layers." This mechanism allows the model to focus on critical parts of the input data when predicting outcomes. This way, LLMs can maintain thematic and stylistic consistency across longer text segments while subtly adapting their outputs to the context.

Continuous Learning and Adaptation

Unlike traditional LLMs, which train on static datasets, some LLMs learn continually by incorporating new inputs they receive while in operation. This way, they can continue to evolve with time or in new circumstances and respond to the changing nature of language and its use.

LLMs as a Breakthrough

The emergence of large language models like GPT-4o by OpenAI is one of the most significant achievements in artificial intelligence, particularly in natural language processing (NLP). This breakthrough is marked by several key advancements that set it apart from earlier models, highlighting its game-changing potential (*The Role of LLMs in AI Innovation,* n.d.).

Scalability and Learning Capacity

Current LLMs, including GPT-4o, employ more advanced neural network architectures than past models, with hundreds of billions of parameters trained on massive and varied datasets. With this scale, their capacity to store and retrieve information has increased tremendously, alongside their ability to discern subtleties and complexities that may not be obvious to smaller models. The vast array of data these models train on covers various linguistic structures, idioms, jargon, and colloquial linguistic styles from different languages. This enables a rich understanding of worldwide linguistic diversity (Mearian, 2023).

Impact on Commercial Applications and Research

In commercial applications, LLMs now perform previously impossible functions, powering chatbots and virtual assistants capable of tackling complex customer issues with the ability to understand and empathize with customers. The role of LLMs in research is to wade through and synthesize vast quantities of scientific text, extract data, summarize research papers, and even develop new research ideas.

Ethical and Technical Innovations

LLM development has also prompted an open discussion about ethics in AI, especially in the areas of bias, fairness, and transparency that we have touched on in the context of generative AI. LLMs train using data created by human authors, which can lead to the internalization of human biases. Resolving some of these issues has been a source of experimentation in the training of algorithms, where there is now a greater emphasis on the development of ethical AI, transparent operations, and methods for debiasing AI systems and making their decision-making processes more transparent and understandable.

Democratization of Access to AI

LLMs have played a pivotal role in making AI technologies more accessible, particularly through their use in generative AI systems. By leveraging natural language processing, LLMs enable developers, businesses, and non-experts to harness AI's power without extensive knowledge of IT.

This democratization is further supported by their integration with APIs and cloud platforms, which offer the scalability, versatility, and real-time text processing needed for NLP technologies. This accessibility facilitates broader use, from sophisticated AI applications to simpler NLP models in smart homes that control IoT devices. As LLMs and NLP technologies continue to evolve, they will likely deepen AI's integration into daily life, making AI tools more intuitive and accessible to all (How LLMs became the cornerstone of modern AI, 2023).

How an LLM "Understands" and Replicates

Large Language Models based on transformer architecture mark a big step forward in language processing by machines. Complicated deep machine learning techniques are the main principle underlying their capacity to "understand" and replicate human language (Do Large Language Models Know What They Are Talking About? 2023).

Transformer Architecture

The transformer architecture presented by Vaswani et al. (2017) is a central element of contemporary LLMs. This architecture allows us to move away from the sequential data processing performed by RNNs and LSTMs and incorporate a system that processes all words simultaneously. This parallel processing ability is key to the transformer model's holistic, contextual understanding of different linguistic data features such as word meanings, syntax, and the relationships between words in a sentence.

Attention Mechanisms

At the core of the transformer model's innovation lies its attention mechanism. This feature enables the model to selectively focus on different parts of an input sentence. For instance, the word order in a sentence does not affect the model's ability to pay attention to relevant words that appeared earlier in the text. This capability to consider the meaning of each word both individually and in relation to other words in the sequence allows the model to generate more coherent and contextually appropriate responses.

The "understanding" of an LLM has no semantic comprehension. So, unlike people, their reasoning is confined to patterns learned from data used during training. They are not conscious and can produce illogical or biased content if these patterns exist in their training data (Acerbi & Stubbersfield, 2023).

To sum up, LLMs "understand" and mimic language using scalable algorithms that can process and generate text with the same patterns they learned from large datasets. These models' capacity to study the entire context of a text enables them to carry out language-centered tasks effectively, taking natural language processing to another level.

The Evolution of LLMs

The emergence of LLMs represents a dramatic developmental leap in AI technology, particularly in the branch of natural language processing. This evolution has developed from basic prototypes that could only perform simple tasks to the emergence of highly advanced systems that can meaningfully respond to user inputs, translate between languages, compose artistic work, and perform many other advanced creative tasks (The history, timeline, and future of LLMs, 2023).

Early Language Models

Early language models were rule-based systems that relied on expert-crafted linguistic rules. Due to their dependence on pre-programmed rules, they struggled to scale and adapt to the complexities of human language. The advent of statistical models marked a significant shift, leading to the development of so-called "n-gram models." These models use large amounts of textual data to calculate probabilities, learning to predict words based on the likelihood of their occurrence following a sequence of preceding words.

Introduction to Neural Networks

The application of neural networks in language modeling, especially recurrent neural networks (RNNs) and their variants like long short-term memory networks (LSTM), has brought further change and improvement to language modeling. These networks proved efficient in working on sequences, making them suitable for tasks that involved understanding contexts throughout long texts. Nevertheless, they were still limited by needing help understanding extended contexts due to training challenges and the vanishing gradient problem. The latter occurs during the training of deep neural networks when gradients used to update the model's weights become very small, causing the learning process to slow down or stop because the model's parameters barely change with each update.

The transformer architecture we mentioned earlier is a milestone achievement in NLP that uses neural networks. Transformers also use self-attention to weigh the importance of each word in a sentence, irrespective of its position. They can capture dependencies between distant words without relying on sequential processing, advancing them over simple neural networks (Shukla, 2023).

Breakthrough Models: BERT and GPT

BERT (Bidirectional Encoder Representations from Transformers): Developed by Google, BERT introduced a new approach to pre-training language models. By using bidirectional training, BERT can simultaneously learn context from both the left and right sides of a word in a sentence and understand the full context of each word based on its surrounding words, leading to a better overall comprehension of the text than would be possible with linear dependency.

GPT (Generative Pre-trained Transformer): The GPT series builds on the transformer architecture introduced by OpenAI, which is continually scaling. With each iteration of the GPT series, from GPT-1 to GPT-4o, the models' data training parameters and abilities increase. They are adept at understanding context, creating natural language text, writing essays and poems, and engaging in user dialogue.

The Use Cases and Future Potential of LLMs

LLMs have rapidly transitioned from mere experimental niche technologies to core components in many applications across different sectors. Their ability to generate natural language dialogue and universal applications make them key technologies for multiple industries and tasks. In the following section, we will examine the current functions of LLMs and how they can affect future implementations and developments.

Current Uses of LLMs

Customer Service and Support: LLMs enable chatbots and virtual assistants to understand and respond to customer requests with a very high level of accuracy. Such AI systems can handle multiple customer service channels simultaneously and provide quick replies to FAQs, thus allowing human agents more time to deal with complex queries (Singh, 2023).

Content Creation and Journalism: Media organizations and producers use LLMs to automate routine reporting, such as sports summaries and financial updates, and create creative content, increasing productivity and allowing human journalists and writers to spend time on in-depth storytelling and analysis.

Language Translation: LLMs produce more accurate and contextually relevant translations than previous models, breaking language barriers to simplify communication globally.

Educational Tools: In the education system, LLMs teach through tutoring, providing examples, generating practice quizzes, and giving feedback on essays and assignments.

Legal and Healthcare Documentation: In the legal and healthcare sectors, professionals use LLMs to draft documents, summarize case laws, and manage patient records to save time and reduce the risk of human error.

Programming Assistants: Developers use tools like GitHub's Copilot, which provides code suggestions and functions based on user comments and coding patterns, to considerably speed up the coding process.

The Future of LLMs

Personalized Interactions: While LLMs today already offer some customization based on user input, future LLMs will deliver exponentially more personalized experiences. Leveraging deeper insights into individual interests, history, and context will provide highly tailored responses, significantly boosting user engagement and satisfaction.

Deeper Contextual Understanding: Current LLMs can understand context and basic sentiment, but future advancements will enable them to grasp even more nuanced elements of human communication, including recognizing emotional subtleties and underlying tones in conversations. This will lead to interactions that feel truly empathetic and human.

Integration across Devices and Platforms: Although LLMs already power various devices and platforms, the future will see seamless and more sophisticated incorporation. LLMs will operate more fluidly across smartphones, wearable tech, and IoT devices, making digital interactions increasingly lifelike and contextually aware.

Advancements in Multimodal Capabilities: LLMs are integrating text with other media formats. In the future, they will excel at combining and synthesizing information from video, images, and audio, providing a richer and more immersive digital experience beyond current capabilities.

Ethical AI and Bias Mitigation: Ethics and fairness will determine the future of LLMs. Strategies must be created to guarantee that LLM outputs are non-discriminatory, objective, and transparent. Privacy and data security must also be respected.

Regulatory and Ethical Frameworks: With LLMs becoming more capable and widely used, a legal framework will be required to regulate their application, especially in sensitive cases. Creating transparency, accountability, and user data protection standards will be essential. We will discuss these topics in more detail in the final chapter.

How LLMs Impact Information Consumption

Large Language Models dramatically alter how we interact with, process, and comprehend information. This trend will affect all areas of consumer information, from education to service sectors.

Accelerating Information Retrieval

LLMs improve search operations by analyzing and processing queries written in natural language much better. Users can ask questions in simple language and will likely get helpful responses immediately. LLMs are superior to traditional keyword-based search methods due to the new models' ability to interpret a query's intent. This results in more accurate and contextually appropriate results, shortens the time for processing unimportant data, and speeds up the information extraction process.

Summarization and Digestion of Information

One of the greatest contributions of LLMs is the speedy digestion and summarization of lengthy texts. This feature plays a significant role in law and academia, where professionals usually handle vast information. LLMs can give short summaries of long research papers, legal documents, books, or reports, allowing readers to pick out the main ideas instead of skimming through the entire text, saving time and making information easier to access.

Enhancing Understanding of Complex Topics

LLMs excel at breaking down complex fields into simple explanations, which is very important in educational contexts. They can explain difficult concepts simply, adjust explanations to suit a student's level, and provide more detailed answers to their questions. The presence of LLMs as digital tutors or mentors improves learning and enhances the concept of personalized education by adjusting to the particular learning styles and paces of different students.

Multilingual Access to Information

LLMs are changing the way information is consumed by removing language barriers. Sophisticated language models do high-precision translation and produce content that can be understood by people speaking different languages, making information available to a wide range of people. This multilingual capability enables people of all cultural backgrounds to access diverse information in their native language, thereby enabling inclusivity and diversity.

Creation of Interactive and Dynamic Content

LLMs lead to more engaging and dynamic types of content based on user responses or choices. In educational software, LLMs can develop problems with increasing difficulty levels according to a learner's performance. This interaction provides a more exciting and personalized information consumption experience.

ChatGPT: A Case Study in Human-AI Interaction

Introducing ChatGPT

ChatGPT is a revolutionary invention from OpenAI, a landmark in the advancement of conversational artificial intelligence. As an extension of the Generative Pre-trained Transformer (GPT) architecture, ChatGPT interacts with users and understands and produces text in a natural and contextually relevant way. Let's consider how ChatGPT works and why it embodies a new paradigm in human-AI interaction (Know it all: ChatGPT and its key capabilities, n.d.):

Built with GPT Architecture

Transformer-Based Model: The innovative transformer model is at the heart of ChatGPT. Unlike most models that work sequentially through words, it uses self-attention mechanisms to evaluate words based on their context throughout the sequence, which improves the model's understanding and ability to generate appropriate responses.

Pre-training Using Diverse Data: ChatGPT initially trains on a large corpus of data obtained from books, websites, and other texts on the internet. This extensive training phase endows the model with a general knowledge of language mastery, grammar, facts, and user interactions in different scenarios and domains.

Fine-Tuning and Reinforcement Learning

Supervised Fine-Tuning: After the primary pre-training, there is a stage called fine-tuning, in which the various ChatGPT applications train on more targeted datasets appropriate for the tasks they will complete. For example, if ChatGPT works as a customer care bot, it would be trained with data from customer care conversations. In this supervised learning phase, the model practices producing correct outputs based on specific inputs to minimize errors.

Reinforcement Learning from Human Feedback (RLHF): Reinforcement learning from human feedback is also applied to ChatGPT to enhance its features. Human trainers give feedback on the model's responses to several scenarios, thereby "reinforcing" it when it produces helpful, accurate, or appropriate outputs. This method helps the AI learn from input data and real-life use scenarios.

Enhanced Conversational Abilities and Contextual Comprehension

Contextual Comprehension: Contextual awareness refers to the identification of words and the interpretation of questions or remarks. This ability enables the model to track topics that contain subtle, non-explicit, or multilevel meanings and produce more natural conversational responses.

Generating Human-like Text: The model's generative nature enables it to create complete sentences and even entire paragraphs that respond to users' questions and interact with them with natural-sounding responses. Models can return appropriate tones, styles, and even humor as may be required depending on the interaction taking place (Ortiz, 2024b).

Significance in Human-AI Interaction

ChatGPT represents a major step in the development of AI communication, thanks to its use of GPT deep learning and fine-tuned reinforcement learning technologies that result in more effective communication. This model can learn about user preferences and tendencies over time, which makes it an essential tool in many sectors, such as education, customer relations, and content development.

ChatGPT not only correctly interprets and responds to messages it receives but also actively learns from user interactions on an ongoing basis, evolving as a system capable of handling the complexities and challenges characteristic of natural human communication and adjusting to individual users' expectations.

Popular and Effective Uses of ChatGPT

Thanks to comprehensive features that improve its efficiency and interactions, ChatGPT has quickly found application in multiple fields.

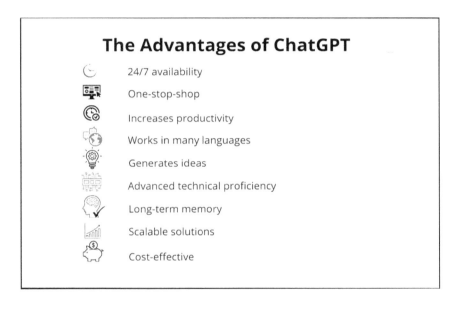

Fig. 2: The Advantages of ChatGPT

EDUCATION

Tutoring and Assistance

Interactive Learning: ChatGPT connects with students one-on-one and can help demystify certain subjects. Its ability to respond to questions in real time makes it a valuable learning tool offering immediate academic support.

Personalization: Based on previous chat interactions, ChatGPT can track a student's performance and learning progress and create tailored teaching techniques to match specific learning speeds and approaches, making the process exciting and efficient and meeting students' individual needs while it develops their talents (Montenegro-Rueda et al., 2023).

Content Generation

Curriculum Development: Educators can use ChatGPT to develop individualized teaching aids and learning approaches. They can also create lesson plans, activities, and teaching resources that reflect their curriculum goals and educational standards.

Assessment Creation: ChatGPT can create various assessments, including quizzes, tests, and assignments. It can generate questions and problems that challenge any level of understanding, helping educators gauge their learners' progress more effectively (Dempere et al., 2023).

CUSTOMER SERVICE

Automated Responses

Efficiency in Interaction: ChatGPT has a high capacity and can manage multiple requests simultaneously, significantly improving response times for teams handling inquiries. This capability is crucial during periods of high customer influx or sales events (ChatGPT for customer service: Limitations, capabilities, and prompts, 2024).

Consistency and Accuracy: By integrating ChatGPT into their applications, companies can ensure they receive consistent information about their customers and guarantee high-quality assistance, preventing discrepancies that can arise from different workers providing varied responses.

Personalization

Enhanced Customer Engagement: Drawing on its accumulated interaction data, ChatGPT can address customers by name, refer to their previous transactions, and make suggestions based on prior conversations. This kind of individualized attention can increase customer retention and satisfaction.

Adaptive Responses: One of ChatGPT's particular strengths is its ability to adapt its language to a customer's mood and tone and the nature of their questions, which is vital for carefully handling delicate or complicated questions (Dilmegani, 2024a).

CONTENT CREATION

Copywriting and Blogging

Scalability: Content creators and marketers can apply ChatGPT to create relevant and high-quality content quickly. It can rapidly generate drafts, social media posts, and marketing copy so that teams can concentrate on other aspects of their work, like strategy and project design.

SEO Enhancement: By incorporating SEO keywords and creating content that caters to audience interest, ChatGPT produces engaging articles and blogs that are SEO-friendly.

Scriptwriting and Ideation

Creative Assistance: ChatGPT can aid scriptwriters by suggesting plots, character dialogues, and scene settings. This support is extremely useful for brainstorming and can spur new creative approaches.

Time Efficiency: Advertisers can use ChatGPT to write scripts for commercials, videos, and presentations and can save considerable time transforming ideas into final scripts (Welance, 2023).

As these applications demonstrate, the model can optimize tasks and transform operations in the educational, customer, and content industries. As AI technology advances and develops, the application of ChatGPT in these sectors is likely to intensify in the future, expanding human-AI synergies.

ChatGPT Makes AI More Accessible

ChatGPT is designed to make AI accessible on a mass scale, opening up revolutionary new possibilities for users and companies across many fields and industries.

Lowering Barriers to Entry

Ease of Use: Any user can easily interact with ChatGPT without advanced programming skills or detailed knowledge of AI's workings. Its simple user interface enables individuals and companies to incorporate complex AI features into their activities without needing professional training or specialized personnel.

Wide Applications: There are numerous possibilities for applying ChatGPT technology, such as assisting in preparing lessons and assignments for students or corporations that use it for customer support and copywriting. Its customizable nature makes it flexible enough to be applied in different fields, thus making it a valuable tool for various businesses (How is AI tech like ChatGPT improving digital accessibility? n.d.).

Democratizing AI Technology

Cost-Effectiveness: By reducing the costs of developing and maintaining AI systems, OpenAI provides a valuable and widely accessible AI tool online, enabling small businesses and individuals to benefit from advanced technologies like large companies.

Innovation and Creativity: ChatGPT encourages users to think outside the box and try out AI in various areas, such as writing, artwork, and problem-solving. This broad accessibility supports diverse AI applications and a wide variety of users, thereby maximizing novel ideas and insights.

The Limitations of ChatGPT

Context Retention: In its first iterations, ChatGPT tended to "forget" the interaction context of previous messages during more extended conversations. This restriction could sometimes result in off-track answers or unsuitable solutions for users. However, thanks to a recently introduced feature, ChatGPT can retain its "memory" across its various chats with a user and draw on these to provide better replies (13 ChatGPT limitations that you need to know, 2024).

Understanding Nuance: Some aspects of communication, such as irony, humor, and references to specific cultural phenomena, may be challenging for ChatGPT to grasp. These aspects are quite complex for AI because they are highly contextual and often subjective. However, ChatGPT can provide more appropriate answers that target the mentality, cultural and semantic norms, and even legal or social aspects of a particular society in the language of the culture the user is interested in. It can also tell jokes if asked to do so.

ChatGPT and Misinformation

The generative capabilities of ChatGPT pose risks related to disseminating misinformation and challenges concerning the originality of content.

Spread of Misinformation

Realistic Yet Inaccurate Content: Because ChatGPT learns and responds based on the patterns it learns in training, the model can generate plausible but false or misleading information, which can be especially critical for domains such as news, education, and scientific discourse, where truthfulness is paramount.

Management and Oversight: To avoid these risks, particularly in highly sensitive environments, it is essential to monitor, fact-check, and occasionally correct ChatGPT on a needs basis to ensure that it does not spread incorrect information (Tucker, 2024).

Content Authenticity

Distinguishing AI from Human Content: Since using AI to produce articles is already a reality, it could become challenging to differentiate between content authored by AI and humans, which would create an ethical dilemma regarding the authenticity and reliability of information across various platforms. This could be especially critical in areas such as academia and journalism.

Ethical Considerations: The challenge includes utilizing AI competently, ensuring transparency, informing users about their interactions with AI products, and addressing issues of consent, privacy, and the responsible use of AI technologies.

Key Takeaways

Large Language Models (LLM) are artificial intelligence systems trained to generate and interpret human language using deep learning techniques based on huge volumes of textual data.

- **Sophisticated neural networks** containing billions of parameters power LLMs. They train on a vast array of texts covering a wide range of styles to mimic different language patterns and enable the model's differentiated recognition capabilities.

- **Unsupervised learning techniques** are used to train LLMs to enhance their language prediction capacity and ability to recognize patterns and correlations within data without formal instruction.

- **Predictive Capacity:** The models' ability to predict what comes next in a sequence enables LLMs to write fluent and contextually relevant texts and complete complicated sentences or documents using natural-sounding language, often indistinguishable from texts penned by human authors.

- **Attention mechanisms** in recent AI models ensure thematic and stylistic cohesion in longer texts and consider subtle contextual signals in responses.

- **Remarkably Versatile:** By providing valuable capabilities for advanced chatbots, content creation, customer support, and research, LLMs have proven the range of their applicability in business and academic environments.

Large Language Models represent a significant advancement in artificial intelligence's ability to understand and generate human discourse. As these models continue to develop, they seamlessly integrate into daily technology use, transforming how people interact, study, and acquire knowledge. With this evolution, it is crucial to consider the ethical implications and strive for responsible use. By doing so, we can harness the capabilities of LLMs to enhance AI accessibility and stimulate new applications in human-machine interactions.

3

THE POWER OF DEEP LEARNING

D eep learning, a type of machine learning in which neural networks can learn from unstructured or unlabeled data, has gradually emerged as the fundamental technology behind several key inventions across different fields. This chapter explores how deep learning technologies have revolutionized specific industries, notably the health sector, automobile industry, security, and commerce and industry. In it, we will examine how these technologies improve productivity and create and redefine opportunities that may determine the future of these industries.

Deep Learning Demystified

Understanding Deep Learning: Mimicking the Brain

Deep learning is an innovative part of machine learning created to mimic the structure and functioning of a human brain when it comes to receiving, analyzing, learning from data, and drawing conclusions. Artificial neural networks based on the biological neural networks in the brain execute this work. These networks allow machines to understand and categorize data in multiple ways. Next, we will explore how deep learning works and some parallels to human neural functions (Introduction to deep learning, 2024).

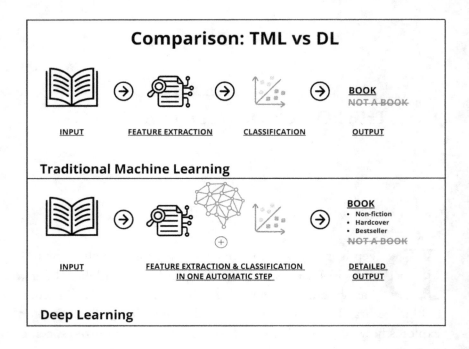

Fig. 3: Traditional Machine Learning vs Deep Learning

NEURAL NETWORKS: THE CORE OF DEEP LEARNING

The Architecture of Neural Networks

Layers of Neurons: We have seen that deep learning architectures consist of layers of nodes, known as "neurons," designed to perform specific input transformations governed by weights and parameters learned from data. These transformations produce valuable outputs, like predictions or classifications.

Depth of Layers: The primary difference between deep learning and more conventional neural network architectures is the depth of the layers. The multiple stacked layers of deep neural networks make it possible for an AI model to learn higher-level features at each layer. An initial layer may learn to detect edges in an image, a subsequent layer to recognize textures, and the following layers to identify patterns and specific object features (Everything you need to know about deep learning: The technology that mimics the human brain, 2022).

How Neural Layers Function

Data Processing: Every neuron in a layer assigns a weighted sum to the data it receives and then applies a nonlinearity before moving to the next layer. This configuration allows the network to integrate the features learned in the previous layers to develop more complex representations.

Representation Learning: By using weights and applying nonlinear transformations, deep neural networks learn to represent the world so that, with little abstraction from the input data, they can solve specific problems like image classification and speech recognition.

LEARNING PROCESS: HOW DEEP LEARNING MODELS SELF-UPDATE

Forward Propagation

Data Flow: A neural network receives input data, such as images or sentences, during forward propagation. This data moves through multiple layers of the network, processing and transforming at each layer until it reaches the final layer and produces a prediction or output.

Activation Functions: While forward propagation enables a model to analyze the edges, colors, and textures of an image at various neural levels, non-linear activation functions at each neuron allow the network to learn and model complex patterns, such as differentiating between images of a cat and a dog. This capacity is crucial because most real-world data is non-linear (Heller, 2019).

Backpropagation and Optimization

Error Correction: Backpropagation is the primary process through which deep learning networks learn. After the network has made a "prediction" (attempted to recognize a pattern or feature), it calculates its errors by comparing the predicted output to the true output (labeled material from training).

Adjusting Weights: This error is passed back through the network and used to adjust its weights and biases, improving its pattern recognition over time. Deep networks can learn complex feature hierarchies, enabling them to identify new examples by combining learned features. For instance, once a model learns to recognize the essential features of the number "3," it can identify it in various contexts, regardless of color, texture, or other feature elements.

Mimicking Neuroplasticity

Neuroplasticity Analogy: Neuroplasticity is the process through which the brain alters its structure and function by creating new neural connections with new experiences. Likewise, neural networks adapt to new data in the same way that their inner parameters or weights adjust for better performance. This adaptation enhances their performance in subsequent iterations and results in learning and memory similar to human cognition (Baheti, 2021).

Not only does deep learning imitate the layered architecture of the brain, but it also uses a systematic approach to modify and optimize the connections in its neural network with the help of feedback, bringing to light a technique that has its roots in biology and applying it within the field of machine learning. This capability allows deep learning models to solve a multitude of problems more effectively as they train on new data in a process very similar to how humans learn, marking one of the significant pillars of training modern AI systems.

Biological Neuroplasticity: A Quick Guide

The similarities between deep learning in AI and human neurological functions merit a brief detour for a closer look. The term "biological neuroplasticity" is defined as the ability of the brain to change its structure and function at different phases in an individual's life. This flexibility is due to a number of reasons, such as learning, experience, and even injury. The link between biological neuroplasticity and how deep learning models work can help us understand how AI systems change and develop (Cheng, 2024).

UNDERSTANDING BIOLOGICAL NEUROPLASTICITY

Mechanisms of Neuroplasticity

Synaptic Plasticity: A key concept in neuroplasticity is that the brain can change the strength of the connections between neurons depending on the frequency and properties of neural activity. This change in the synaptic connections enables the strengthening or weakening of these pathways and plays an essential role in learning and memory.

Structural Plasticity: In addition to synaptic modifications, the brain may undergo structural changes, such as generating new neurons or synapses, pruning older ones, or reorganizing the internal architecture of neurons. These changes can occur during learning spikes or after a recovery period from a head injury.

Factors Influencing Neuroplasticity

Stimulation and Learning: New experiences that stimulate the brain, such as picking up new hobbies or learning new skills, cause changes in its structure.

Injury and Recovery: When part of the brain is damaged, it can recover or reroute its functions to other neural pathways, thereby proving its plasticity.

The Analogy of Biological Neuroplasticity in AI: Adaptive Neural Networks

In artificial intelligence, especially deep learning, the ability of neural networks to adapt and optimize their structure and configurations in response to data inputs mirrors biological neuroplasticity.

Weight Adjustment and Learning

Training and Adaptation: In deep learning, a neural network updates its weights via backpropagation, where the network tries to rectify the difference between the expected output and the real output it has provided. This method is similar to a feedback mechanism in the human brain, which either strengthens or weakens the synaptic connection.

Continuous Learning: Some neural networks learn new information incrementally, meaning they can update their knowledge base with new data without being retrained from the ground up. Their learning is continuous and lifelong, a characteristic also observed in human brains.

Applications of Adaptive Networks

Real-time Data Processing: In systems where data input is constant and diverse, adaptive neural networks work best because they modify their parameters to better agree with the most recent data, just as people change their perceptions of reality.

Predictive Analytics: These types of networks can develop their prediction capabilities over time as they receive new data, making them perfect for use in finance, healthcare, and all the other fields where prediction is a complex process involving many variables.

Challenges and Opportunities

Complexity of Adaptation: Mimicking biological neuroplasticity in AI involves intricate algorithms and significant computational power, presenting challenges to be addressed and development opportunities.

Ethical Considerations: As these networks become more flexible, identifying and excluding risk factors in a model's learning process, such as biased training data and the need for transparency in its decision-making procedures, become more significant. An analogy would be ensuring that a judge is objective and without prejudice (Nosta, 2023).

Key Technologies in Deep Learning

Thanks to its particular technologies and approaches, deep learning is effective in numerous fields and widely used in modern artificial intelligence systems. As we mentioned earlier, the two main elements of artificial neural networks are Convolutional Neural Networks and Recurrent Neural Networks. To understand deep learning in AI, we will examine these and other deep learning technologies in more detail now and review their applications (Takyar, n.d.).

CONVOLUTIONAL NEURAL NETWORKS

Architecture and Functionality

In chapter one, we touched on Convolutional Neural Networks (CNNs), a type of neural network well-suited for a data structure laid out in a grid, such as identifying images. CNNs use a mathematical operation known as convolution, which is crucial in their input transformations. It is helpful to look at a brief breakdown of the architecture and functionality of CNNs, including the principles that govern their operations.

CNNs have several layers, including convolutional, pooling, and fully connected layers. Convolutional layers use filters to identify the main features of input data (edges, textures, patterns, etc.). Afterward, pooling layers downsample the feature maps, retaining important information while reducing dimensionality. The fully connected layer produces the output (what the model "sees" in the image) (Matsuo et al., 2022).

Convolutional Layers:

In convolutional layers, the input data, such as an image, is passed through a set of learnable kernels known as convolutional filters. These filters help the AI model recognize and activate specific features within the data.

A functional analogy can be helpful: Imagine you have a large painting and want to find specific flower patterns. Instead of examining every detail, you use a small stencil (filter) with a cut-out of the flower shape and slide it across the painting (convolution). You mark those spots on a new canvas (feature map) each time the pattern matches. Using different stencils for various patterns, you create multiple feature maps.

In CNNs, filters slide over images to detect patterns, and feature maps record where these patterns appear, enabling the network to recognize complex shapes and objects. The network learns filters that fire when they detect certain features at specific locations in the input, such as edges or particular textures.

Pooling Layers:

Pooling layers help reduce the size of the feature maps in a neural network while preserving the most important information. This process, known as downsampling or subsampling, simplifies the data, making the network more efficient.

There are two types of pooling. Max pooling examines small sections of the feature map and retains only the highest value from each section. These "values" are represented by pixels or feature activations that highlight detected colors, shapes, textures, and other patterns within the image. Thus, max pooling effectively reduces the size of the feature map by half.

On the other hand, average pooling calculates the average value of each small section of the feature map, reducing the size similarly to max pooling. By using these techniques, CNNs can reduce the dimensionality of feature maps while preserving important information, making the model more efficient and robust in recognizing patterns and features within the input data.

Pooling layers are crucial because they reduce the number of parameters and computations the model needs. Additionally, pooling improves the model's recognition of features regardless of their size or orientation in the image. AI can more reliably identify objects under different conditions, making it more useful in real-world applications like image and video recognition.

While pooling layers are most commonly used in image recognition tasks, we can apply their principles to domains like natural language processing, audio processing, and time-series analysis. They help reduce data dimensionality, extract essential features, and make models more robust and efficient across different data types. For example, max pooling can capture the most significant words or phrases that indicate sentiment in sentiment analysis. In speech recognition, pooling layers distill key features from audio signals to improve recognition accuracy. In anomaly detection, pooling helps highlight significant deviations in time-series data. Finally, in technical analysis for financial markets, pooling can summarize critical trends and patterns from stock price movements.

Fully Connected Layers:

The last layers of convolutional neural networks consist of fully connected layers, which follow several iterations of convolutional and pooling layers. In a fully connected layer, each neuron connects to every neuron in the previous layer. The final fully connected layer consolidates the features learned in the previous layers to make decisions on the inputs, thereby generating an overall interpretation of the image by the model.

EXAMPLES OF CNN APPLICATIONS

Image and Video Recognition

- **Security and Surveillance:** CNNs are essential in security systems for facial recognition to improve security measures in areas where security is paramount, like airports and banks. They parse faces from videos to recognize people.

- **Smartphone Security:** Modern smartphones, for example, use CNN-based facial recognition as a method of biometric authentication to unlock a device or confirm a transaction.

- **Retail:** CNNs are used for cashier-less checkouts in retail spaces. They can recognize products placed in a cart and count goods without scanning barcodes.

- **Safety Monitoring:** At workplaces, CNNs analyze camera feeds to identify potentially risky actions or people in forbidden zones and improve safety measures.

- **Tourism and Marketing:** CNNs examine scenic images shared on social media to pinpoint famous landmarks, aiding tourism boards and marketeers in updating and optimizing their content.

- **Content Categorization:** Media companies also use CNNs for automated scene and object tagging, and for updating and optimizing their content.

Medical Image Analysis

- **Tumor Detection in Radiology:** CNNs interpret MRI scans and other medical images to identify and outline tumors. For instance, they can differentiate between malignant and benign tumors by the shape, size, and texture of the growths.

- **Diabetic Retinopathy:** CNNs identify the patterns of retinal images to diagnose diabetic retinopathy, which may cause blindness if not treated. They assist ophthalmologists in diagnosing the disease at an early stage, which enhances its management and treatment.

- **Organ Delineation:** In treatment planning, especially in radiation oncology, CNNs use image segmentation to define the organs and tumor volumes for targeted radiation delivery and limit the effect on healthy tissue.

Autonomous Vehicles

- **Road Safety:** CNNs are instrumental in helping self-driving vehicles detect different objects on the road, including cars, people, bicycles, and signs. This ability is essential for making critical decisions in real time.

- **Lane Detection:** CNNs assess the real-time video stream on the road to recognize lanes and control a vehicle's position by adjusting the steering.

- **Traffic Prediction and Management:** CNNs can estimate traffic conditions using video streams from different areas and inform drivers about the best routes to minimize traffic jams.

- **Weather Condition Adaptation:** CNNs allow autonomous vehicles to adjust their driving style to weather conditions through visual inputs, slowing them down during rain, fog, and snow.

Recurrent Neural Networks

As we recall, RNNs are the other major artificial neural network. Unlike CNNs, they are suitable for processing sequential data. RNNs use a kind of internal "memory" that allows them to process arbitrary inputs, which makes them ideal for language processing or time series analysis.

STRUCTURE AND MECHANISM

Sequential Data Handling

Looping Mechanism: Unlike CNNs, which are unidirectional feed-forward neural networks, RNNs have loops in their architecture, enabling them to retain information across subsequent inputs in a sequence and mimic abilities comparable to "memory."

Processing Sequences: The "neurons" of RNNs build on previous calculations. This can be compared to a sequence of repeating cells in a neural network. At each iteration, the module receives an input from the dataset and the output from the previous iteration.

Challenges with Long Sequences

Vanishing Gradient Problem: Like other types of neural networks, RNNs suffer from the vanishing gradient problem. Gradients update the network's weights during training. However, they shrink when propagated backward through many neural layers or time steps, making it challenging for the RNN to capture and memorize data present in long sequences since the early inputs do not significantly affect the outputs.

We can think of this in terms of an analogy. Imagine you have a very long train with many connected cars. The engine gives the first car a tiny push at the train's front end. This impetus needs to travel all the way to the end of the train. However, as the push travels from one car to the next, it gets increasingly smaller until it can hardly be felt when it reaches the last car.

Effect on Learning: The vanishing gradient problem mainly impacts the network's learning of long-range relationships with extended sequences such as long text documents or time-series data. As a result, a network may learn very slowly or not at all, especially for long-range dependencies, because the early layers receive almost no update signal.

Variants and Improvements

To improve the ability of RNNs to learn long-term dependencies, researchers have proposed improved models to alleviate the vanishing gradient problem. Two of the most popular are Long-Short-Term Memory (LSTM) units and Gated Recurrent Units (GRUs).

Long Short-Term Memory Units

How They Work: LSTMs are a particular recurrent neural network designed to remember important information for long periods and forget irrelevant details, solving the vanishing gradient problem. They work with a special architecture composed of gateways.

Forget gates decide what information to discard from memory. They look at the current input and what the model remembered before and choose what data to keep or discard.

Input gates allow new information to be added to the memory. They decide what aspects of new information are important and update the memory accordingly.

Output gates decide what information from the memory is used for the next step and define the next hidden state—the internal memory of the neural network that captures and stores information about the previous inputs.

These gates, taken together, enable LSTMs to update relevant and discard unnecessary information across long sequences, making them perfect for time-dependent problems like language translation, text generation, and even music learning.

Gated Recurrent Units

GRUs are a variation of LSTMs developed to be easier and faster to train thanks to the incorporation of Forget and Input Gates into a single Update Gate.

- **Update Gates** help the model understand how much old information to erase and how much new information to retain.

- **Reset Gates** determine how much past information should be forgotten and enable the model to prune away any data unnecessary for forecasting.

- **Efficiency and Performance:** Compared to LSTMs, GRUs have a simpler structure with fewer parameters since the two gates are integrated. This reduction in complexity results in faster operation and training times while still maintaining performance, which makes GRUs ideal for scenarios where the memory control of LSTMs may not be essential for gaining superior performance.

Comparative Performance and Applications

GRUs: Examples of use cases in which the superior operative and training speed of GRUs can replace the long-term memory control of LSTMs include generating short texts based on patterns learned from large datasets, translating short texts by generating word sequences, and transcribing spoken language into text.

LSTMs: Applications that require managing long-term dependencies and retaining important information over extended sequences are particularly well-suited for LSTMs. They include time series prediction, such as forecasting stock prices or weather patterns; natural language processing tasks like language modeling; generating long, coherent texts that require advanced memory capability; sentiment analysis; speech synthesis for converting text to natural-sounding speech; anomaly detection in data sequences like network traffic or sensor data; and handwriting recognition, where understanding the sequence of pen strokes is crucial for accurate interpretation.

Both technologies represent a significant advancement over baseline RNNs, offering tools better equipped to deal with the challenges of sequential data.

A CLOSER LOOK AT ADVANCED RNN APPLICATIONS

Speech Recognition

Function: RNNs convert speech to text and are used in various applications, such as voice translation and transcribing speech to text. They work by receiving and analyzing audio signals in short time frames. For each frame, the probabilities of specific phonemes are calculated. These predictions are later used to construct words and complete sentences.

Example: Intelligent personal voice assistants such as Apple's Siri, Google Assistant, and Amazon's Alexa also employ RNN models to interpret voice commands given by users in natural speech.

Language Translation

Mechanism: Within the realm of Machine Translation (MT), RNNs use a sequence-to-sequence model. In this model, one RNN processes the source language sentence and converts it into a summary or representation. A second RNN takes this summary and generates the sentence in the target language.

Advancement: LSTMs and GRUs have been used in services like Google Translate to enhance machine translation quality, thanks to their ability to manage the long dependencies often found in translating words within sentences.

Text Generation

Application: RNNs can generate syntactically coherent text, grasping the internal structures and relationships within large bodies of text. This makes them perfect for applications like intelligent AI chat assistants, crafting complex stories, and creating synthetic text collections.

Real-World Use: Modern models, such as OpenAI's GPT, which generates text using patterns learned from previous texts, build on transformer technology derived from RNN concepts. This allows them to produce coherent, logical, and semantically related content.

Stock Market Forecasting

Usage: Technical analysts use RNNs to develop trading tools that forecast future stock prices based on past prices and other market signals. Their sequential nature makes them ideal for capturing the temporal dependencies inherent in stock price data.

Impact: Financial analysts can use RNNs to make better investment decisions to manage risks and portfolios by predicting sales changes.

Economic Trend Analysis

Application: According to historical data, RNNs predict economic figures such as GDP growth rates, unemployment rates, and inflation. These predictions help inform decision-making regarding economic policies and other necessary provisions.

Benefits: This can enhance the ability of states and economic planning structures to respond to new economic trends, improving their future responses to emerging economic shocks.

Weather Forecasting

Functionality: RNNs forecast long sequences of meteorological data to predict future weather patterns. They excel at characterizing highly variable and nonlinear weather features over time.

Advantage: Improved accuracy in predicting future weather conditions can aid in responding to natural disasters, managing agriculture, and planning everyday activities for individuals and businesses.

Music Generation

Process: RNNs can identify patterns in music tracks from different composers or genres and then create new compositions that follow these patterns. They learn from sequences of musical notes or other representations of music data.

Example: Many applications that provide background music for videos or digital games also use RNNs to create music that corresponds to the content of the game or video.

Video Frame Analysis

Use case: In video processing, RNNs can predict future frames in a video sequence or analyze activities by capturing temporal features, which is particularly important in applications such as surveillance, where monitoring activities over time is essential.

Function: Security cameras in public places or home safety systems can use RNNs to analyze video streams and learn and identify suspicious movements or actions, thus enhancing security without constant human intervention.

Generative Adversarial Networks

DUAL-STRUCTURE FRAMEWORK

Generator and Discriminator Dynamics

Generator: The generator's task is to create data samples that closely resemble real data. It learns to produce realistic outputs from random noise and continually improves its ability by training against the discriminator.

Discriminator: The discriminator acts as a judge in the GAN architecture, evaluating both real and synthetic data created by the generator. Its goal is to distinguish between the two sources accurately. During training, the discriminator becomes increasingly effective at identifying real versus synthesized data.

Training Process

GANs consist of two neural networks trained adversarially, each improving as they compete. Like a minimax two-player game, the generator and discriminator adjust their parameters based on each other's performance. The generator aims to reduce the discriminator's accuracy in distinguishing real from fake data, while the discriminator strives to improve its accuracy.

USE OF GANS IN APPLICATIONS

Data Augmentation

Training Data Enhancement: GANs are particularly useful when training data is scarce, and more data is needed to train accurate machine learning models. For example, in medical imaging, GANs can generate additional labeled images to train diagnostic models without requiring more real patient data.

Handling Imbalanced Datasets: In cases where certain data classes are sparse (for example, rare diseases), GANs can generate more samples for these underrepresented classes, making the dataset more balanced for machine learning algorithms.

Creative Content Generation

Art and Design: During training, GANs learn the styles of different artists and use this to generate unique new art. This technology has also been adopted in fashion, where GANs create new collections based on their training with fashion data.

Music: While RNNs are great for generating music that follows the logical progression of a sequence, GANs are more adept at producing complete polished and varied musical pieces that closely mimic the style of the training data.

Video Games and Virtual Environments: In video games, GANs generate realistic textures and landscapes, enhancing the game's visual appeal without requiring artists to draw every detail manually.

Marketing and Entertainment

Personalized Content: In advertising, GANs create unique promotional messages that suit the specific customer's preferences. Their ability to customize content can extend to changing aspects of a video advertisement in real time to fit the target audience.

Film Production: In the film industry, GANs are used to automatically create computer-generated imagery (CGI) based on inputs or scripts, potentially saving film companies significant time and money.

Transfer Learning

Transfer learning is an established concept in machine learning. In this approach, the expertise gained in one task is used to optimize the learning in a different but related task. This approach is especially useful in deep learning, as training from scratch may take a lot of computational resources and time.

EFFICIENCY AND FLEXIBILITY

Pre-trained Models

Concept: Transfer learning typically involves using a base model trained on an extensive, general database, such as ImageNet—a vast collection of labeled images for object recognition. The pre-trained model has already learned many valuable patterns and features, which it applies to various tasks beyond the one for which it was initially trained.

Advantages: Fine-tuning a pre-trained model saves time and resources and performs well with limited data. For example, by leveraging the expertise gained from the comprehensive object recognition provided by ImageNet, a model can quickly and efficiently learn to recognize individual categories like plant species using a smaller dataset.

Transfer Learning Applications Across Domains

Cross-domain Utility: Transfer learning excels at reusing models trained in one domain for tasks in another, such as general image processing versus medical image analysis or satellite imagery. This flexibility is valuable when labeled training data is limited or costly.

Examples:

Satellite Image Analysis: Standard images fine-tune models applied to interpret satellite imagery for land cover classification and disaster monitoring tasks.

Medical Diagnostics: Models trained on general images can be fine-tuned to detect specific pathologies in medical scans, such as identifying tumors in MRI images or anomalies in X-rays.

Reducing Barriers and Improving Accessibility

Democratizing AI: Transfer learning has simplified achieving high performance by requiring less data and fewer resources than other AI technologies. This democratization allows small organizations and researchers with limited funding to develop advanced technologies.

Enhanced Innovation: AI solutions developed with transfer learning are prototyped and deployed much faster than training a model from scratch, accelerating innovation cycles and opening up a wider range of possibilities.

Enabling Broader Adoption

Education and Experimentation: Transfer learning allows learners and researchers to experiment with advanced models without requiring extensive infrastructure, supporting educational and experimental purposes in AI.

Global Impact: In regions with limited data or computational power, transfer learning empowers local talent to address regional issues, such as natural language processing for regional languages or optimizing local farming practices.

Traditional Machine Learning vs. Deep Learning

Aspect	Traditional ML	Deep Learning
Model Architecture	Uses simpler models like regression, decision trees, and SVMs for classification (e.g., spam detection).	Utilizes deep neural networks (CNNs, RNNs) for more complex tasks.
Data Requirements	Works well with smaller datasets; often uses manual feature selection.	Requires large datasets to learn features from scratch.
Feature Engineering	Relies on manual feature engineering and expertise.	Features are automatically extracted; no manual engineering is required.
Computational Resources	Less resource-intensive; can use standard CPUs.	Requires significant computational power and often utilizes GPUs.
Training Time	Faster training with simpler models and smaller datasets.	Longer training due to complex architectures and large datasets.
Interpretability	More transparent and easier to interpret (e.g., decision trees).	Often considered "black boxes," making interpretation harder.
Output	Suitable for simple tasks like classification and regression.	Handles complex tasks like image recognition and autonomous driving.
Scalability	Limited scalability with increasing data complexity.	Highly scalable, improving with more data.
Flexibility	Less flexible; requires structured, categorical data.	Handles diverse data types like images, audio, and arrays.
Real-Time Use	Easier real-time deployment due to simpler models.	Not always suitable for real-time use without optimization.
Use Case Examples	Spam detection, customer segmentation, credit scoring.	Autonomous vehicles, speech/image recognition, AI in healthcare.

Table 1: Traditional Machine Learning vs. Deep Learning

Transformative Applications of Deep Learning

Deep learning is now one of the most impactful technologies in many fields. It successfully utilizes massive computational resources and large datasets to resolve problems once considered unsolvable. This section will discuss its uses in various industries, including the benefits and risks.

Transforming Healthcare

AI IN DIAGNOSTICS AND TREATMENT

Imaging Diagnostics

Technology: In medical image analysis, deep learning has become a game changer thanks to convolutional neural networks because these are trained on massive datasets of images and can detect features and defects that are often difficult for the human eye to distinguish (Nachaat, M. & Nachaat, M, 2023).

Applications: Artificial Intelligence systems can use medical images, such as MRI, CT scans, or X-rays to diagnose various diseases, fractures, or abnormalities. These systems are accurate and fast, significantly contributing to early diagnosis and treatment planning.

Benefits: In conditions like cancer, for instance, these AI systems can help reduce the time that elapses from diagnosis to commencing treatment, significantly improving a patient's prognosis.

Personalized Treatment

Data Integration: Deep learning systems gather and process information from multiple sources, including patient records, genetic information, and treatment outcomes. They can use complex calculations to discern various patterns and determine the recommended course of treatment for a specific patient (Chen et al., 2021).

Customization: Therapies can be tailored to individual patient needs, optimizing treatment outcomes and avoiding unwanted side effects. For example, deep learning models in oncology predict how a particular type of cancer will likely respond to a given therapy, thus customizing the treatment cycle (Ahn et al., 2021).

DRUG DISCOVERY AND DEVELOPMENT

Molecular Modeling

Function: Deep learning reduces time spent on the drug discovery process through fast simulation of molecular interactions. These models explain how drug molecules behave at the molecular level to infer their effectiveness (Paul et al., 2021).

Impact: This application also has the advantage of shortening the time and cost of the drug discovery process and decreasing the use of physical trials in the initial phases. Machine learning algorithms can quickly sift through millions of compounds to assess which are most likely to be useful drug candidates for synthesizing and further testing (Qureshi et al., 2023).

Clinical Trials

Optimization: Based on past experiences, AI estimates the results of clinical trials for new drugs to improve the design of these trials. It can choose proper inclusion criteria and suitable medicinal dosages (Chun, 2023).

Participant Selection: AI can enrich the probability of a positive result in clinical trials by selecting appropriate participants according to their genetic and phenotypic data.

TELEMEDICINE AND REMOTE HEALTHCARE

Remote Monitoring

Wearable Technology: Smart wearable devices use AI algorithms to constantly track parameters such as heart rate, blood pressure, and glucose levels. These devices can identify deviations from the norm and alert patients and healthcare workers (Nazarov, 2023).

Preventive Care: Continuous monitoring helps identify signs of disease early, reducing the need for hospital admissions and long-term treatment plans, thereby lowering healthcare costs.

Virtual Health Assistants

Accessibility: Virtual doctors provide medical advice around the clock, making it easier for people to seek treatment, especially those with limited mobility or in areas with restricted access to healthcare.

Functionality: Using natural language processing, AI assistants can interpret questions about symptoms and determine if a doctor's visit is necessary, whether the issue can be addressed at home, or if it constitutes an emergency (Sharma et al., 2023).

ETHICAL CHALLENGES AND PATIENT PRIVACY

Data Privacy

Concerns: The use of deep learning in healthcare is predicated on the large amounts of personal data that deep learning algorithms require, prompting discussions about privacy and data protection. It is crucial to safeguard this data and prevent it from being accessed by unauthorized personnel (Moore, n.d).

Regulations: Adherence to rules and regulations such as HIPAA in the United States and GDPR in Europe is significant. These regulations require robust protection measures for using and disclosing patient health information.

Bias and Fairness

Inherent Biases: To prevent and mitigate bias in AI, it is essential to recognize and address inherent risks. The quality of data and the approach used for training are crucial, as AI-based systems can replicate or even amplify biases present in the training data, leading to skewed or unfair medical advice and treatment recommendations, particularly for minority groups (Farhud & Zokaei, 2021).

Mitigation Strategies: Additional efforts are needed to improve the representativeness of data samples and to enhance algorithms' ability to detect and mitigate biases in decision-making models (Gerke et al., 2020).

Thanks to deep learning, the healthcare industry has new diagnostics, treatment, and drug design possibilities. However, these advancements also bring challenges that must be addressed carefully, particularly regarding ethical issues and security concerns.

The Technology Behind Autonomous Vehicles

Self-driving cars (AVs) may be one of AI's most significant use cases. The central technology behind them is deep learning. It enables autonomous vehicles to understand their environment, detect objects, and make real-time decisions to function effectively in transportation systems and safely interact with other traffic participants. Now we will see how deep learning enables critical aspects of AV technology.

Sensor Fusion

Overview: Autonomous vehicles use several types of sensors, such as cameras, radar, LiDAR (Light Detection and Ranging), ultrasonic sensors, and, in some models, infrared sensors. Each offers essential information about a vehicle's environment (Srivastava, 2024).

Deep Learning's Role: Sensor fusion is a process by which deep learning models analyze and make sense of data from various sources. This enables the vehicle to map its surroundings fully and accurately, which is crucial for successful maneuvering and response.

Computer Vision and Object Detection

Function: Cameras are essential for computer vision because they capture the visual data that deep learning models process. These models identify and recognize objects like other cars and trucks, people, signals, and markings on the road.

Techniques: CNNs interpret image data in autonomous vehicles. They help identify objects, determine their positions, and track their movements, which is crucial for avoiding obstacles and choosing safe driving routes (Lutkevich, 2024).

Real-time Decision Making

Processing: Real-time data provided by the car's sensors, combined with deep learning, allows the vehicle's system to make decisions in milliseconds regarding various environmental conditions, just as a human driver would, such as how and when to accelerate, decelerate, overtake, or take a turn (Schroer, 2024).

Models Used: Predicting the movement of pedestrians and other vehicles requires understanding temporal dynamics, for which recurrent neural networks (RNNs), including LSTMs or GRUs, are applied.

Environmental Understanding

Interpretation of Environment: Beyond object identification, deep learning enables AVs to interpret entire scenes, including distinguishing between urban and rural environments, assessing road conditions, and recognizing adverse weather conditions.

Contextual Awareness: Environmental understanding includes not just perceiving context but comprehending, for instance, the meaning of a hand sign from a pedestrian or the actions of a traffic police officer, which are challenging tasks that require advanced AI systems.

Localization and Navigation

GPS and Maps: AVs use the Global Positioning System (GPS) in conjunction with high-definition maps that contain detailed information about roads, including lanes, traffic signals, and structures. Machine learning algorithms combine this GPS data with real-time environmental data from the vehicle's sensors to determine its precise position and compute destination trajectories.

Simultaneous Localization and Mapping (SLAM): Based on deep learning, SLAM enables vehicles to update their maps in real time and adapt to new or modified conditions without human intervention.

Facial Recognition for Security and Marketing

Enhanced by deep learning, facial recognition technology has continuously advanced and is widely used in security and customized consumer services.

Individual Identification

Crowd Surveillance: Facial recognition models based on deep learning algorithms use extensive datasets of facial photos to identify individuals in large crowds. This capability is essential in areas where security is paramount, such as during events, at airports, and other large gatherings.

Law Enforcement: Facial recognition is used to monitor individuals who are of interest to law enforcement agencies. The technology can instantly match surveillance pictures with databases of known people, which helps in investigations and increases security (Najibi, 2020).

Facility Security

Access Control: Facial recognition systems are used in secure areas to ensure that only authorized individuals can gain access. This technology replaces conventional badges or key cards, offering a superior access control mechanism since duplicating or stealing someone's face is nearly impossible.

Real-time Alerts: Security systems with facial recognition features can send alerts in real-time when unauthorized people are detected, thereby increasing the efficiency of security surveillance (Michalowski, 2022).

Customer Engagement

Personalized Advertising: Some store owners use facial recognition to determine the age and gender of shoppers, enabling targeted marketing. For example, digital signage equipped with a camera can adjust its content based on the age and gender of the person in front of it.

Enhanced Shopping Experience: In high-end shops, facial recognition can help staff recognize returning customers and review their previous purchases to recommend other products they might be interested in.

Service Customization

VIP treatment: Facial recognition technology is also applied in hotels and casinos to welcome VIP guests as soon as they arrive so that the staff can treat them accordingly based on their status. This application not only enhances the satisfaction level of the customers but also their loyalty to the brand.

NATURAL LANGUAGE PROCESSING

Machine Translation

Deep Learning Impact: Deep learning has become instrumental in machine translation, especially the transformer model and BERT. These algorithms learn word dependencies in large text volumes and translate texts with greater fluency and relevance.

Real-World Applications: Machine translation supports platforms like Google Translate, enabling real-time translation that is nearly fluent across multiple languages and facilitates multilingual interaction, commerce, and access to information (University of York, n.d.).

Modern statistical translation systems are based on deep learning approaches, allowing them to significantly improve translation quality by considering sentence context, idiomatic expressions, and complex grammar.

Voice Command Processing

Functionality: Smart devices like Amazon's Alexa, Google Assistant, and Apple's Siri incorporate deep learning to interpret spoken commands. They can convert speech to text, analyze texts to understand users' intentions and commands and perform their requests.

Continuous Learning: These systems learn and adapt to the user's voice, accent, and usage patterns and can, therefore, improve their response accuracy with increasing use (University of York, n.d.).

Application Diversity

Home Automation: Assistants are central to smart home solutions, controlling lights, temperature, security systems, and other aspects through voice commands.

Personal Assistance: Home assistants can help users increase their convenience by setting alarms, playing music, and providing weather forecasts.

Impacts on Business and Finance

AI in Investment and Asset Management

Automated Trading Systems

Deep learning applications that exercise autonomous trades have emerged as critical players, changing how investments and assets are managed through automation and data analysis. Two applications are especially significant: algorithmic trading and portfolio management.

Functionality and Implementation

Data Processing: Computer-based trading systems use sophisticated software to analyze market information in real time, including price, quantity, and trends. These systems can process and interpret data much faster than humans and on a scale unmatched by them.

Trade Execution: AI systems can make trades at very profitable times by acting in real time and taking advantage of minute price movements that a human could not capture (Artificial intelligence: The next frontier in investment management, n.d.).

Benefits

Increased Trading Efficiency: AI systems find the best way to execute a specific order by buying or selling at the right time and price to get the best returns.

Accuracy and Speed: These systems minimize errors, such as incorrect order entries, and respond instantly to market changes.

Continuous Operation: Unlike human traders, AI systems are untiring and can operate 24/7, capturing opportunities in global markets across different time zones (Bartram et al., 2020).

Portfolio Management

Robo-Advisors

Personalized Investment Advice: Robo-advisors collect data on individual investors, including their risk profiles, financial objectives, and changing investment preferences. These platforms use algorithms to create and rebalance portfolios based on market conditions and the investor's needs.

Accessibility and Cost Efficiency: Robo-advisors enhance efficiency, reducing the need for human financial advisors and lowering the cost of investment advice, thereby democratizing and expanding access to wealth management services.

Risk Assessment

Predictive Capabilities: Market prediction involves integrating market trends, economic factors, and geopolitical events into AI models. These forecasts help evaluate and prevent potential losses and maximize gains for various investment strategies.

Scenario Analysis: Sophisticated AI technologies use the scenario approach to determine the impact of various conditions or decisions on investment performance. This helps develop sustainable strategies under different market conditions.

Benefits

Dynamic Portfolio Adjustment: AI-driven systems can actively manage investment portfolios and quickly adapt to market fluctuations. This dynamic rebalancing helps keep portfolios on track for targeted asset allocation.

Enhanced Risk Management: Better insights and predictive analytics allow fund managers to identify and prevent potential risks, which can lead to increased returns at lower losses.

Predictive Analytics in Financial Markets

Integrated predictive analytics powered by AI are crucial to contemporary financial markets, aiding decision-making support and planning. This section explores how AI enhances market trend and sentiment analysis, influencing trading and investment strategies.

Market Trend Analysis

Application and Functionality

Data Integration: AI systems combine past and current information to accurately predict future market trends. These may include stock prices, trading volumes, and economic indicators.

Predictive Modeling: Machine learning algorithms allow AI to capture more intricate relationships and correlations than statistical tools can detect (Sphinx Solutions Pvt. Ltd., 2024).

Techniques

Neural Networks are especially well-suited for use in the financial markets. They can capture non-linear relationships in data, learn from large financial datasets, uncover hidden patterns, and predict future market dynamics.

Ensemble Methods: Algorithms like Random Forest (a collection of decision trees working together) and Gradient Boosting (a step-by-step technique that builds one model at a time) improve prediction accuracy. Combining the strengths of multiple models reduces errors and provides more reliable forecasts, much like consulting a range of experts to get the best advice.

Deep Learning: Advanced techniques such as LSTM networks are well suited for handling time series data, which is common in the financial markets. LSTMs can capture long-term market dependencies, making them ideal for forecasting stock prices or economic developments. (The Impact of Artificial Intelligence in Banking, n.d.).

Benefits

Enhanced Decision-Making: The sophisticated use of AI for trend analysis helps financial analysts and traders choose suitable investments and avoid potential risks by making better market predictions.

Real-time Insights: Traders can quickly make informed decisions based on the extensive data that AI models can analyze in real time.

Sentiment Analysis

Function and Process

Data Sources: AI-powered automated tools can analyze information obtained from financial news, analysts' reports, social networks, and even blogs to measure an audience's attitude toward particular securities or assess market sentiment.

Natural Language Processing: Text analysis and NLP are used to understand and measure the sentiment hidden within textual data. This requires models to recognize subtle language aspects, including positive, neutral, and negative attitudes.

Impact

Market Prediction: Sentiment analysis is a powerful tool because it helps traders recognize people's attitudes, which usually determines market trends. For instance, increasing positive social media sentiment about a new technology product may predict increasing stock prices for the company in question.

Risk Management: Market declines or crashes can be predicted by analyzing changes in trend sentiment. Shifts in public opinion and investor behavior often precede these events.

Challenges

Accuracy and Reliability: The accuracy and reliability of sentiment analysis depend on the quality of its data feeds. Inaccurate information or a bias in reports will influence its predictions.

The complexity of Financial Language: Writing in the financial sector and technical economic reports uses intricate language structures that require advanced natural language processing capabilities for effective sentiment analysis.

The Impact on Traditional Banking

AI is increasingly used in banking to revamp customer experience, operations, and the credit and lending business.

Customer Service Enhancement

Chatbots and Virtual Assistants

Functionality: AI-enabled chatbots and virtual assistants integrated into existing banking services offer basic information and services, such as inquiries about balances, transaction history, or fund transfers. These systems use NLP technology that enables friendly, natural communication with customers.

Benefits: The key advantages include 24/7 availability, rapid responses to customer inquiries, and reduced reliance on human employees. This automation can enhance customer satisfaction through efficient service delivery and lower operational costs while minimizing the need for human intervention.

Personalization

Data Analysis: To personalize banking services, numerous parameters, including a customer's transaction history, browsing history, and previous interaction history, are considered.

Customization: Using customer data, banks can offer tailored product promotions, personalized financial solutions, and targeted messages, significantly enhancing client interest and trust.

Operational Efficiency

Process Automation

Applications: In banks, AI performs back-office tasks like document handling, claims, risks, and compliance checks. For instance, it can scan loan documents and parse relevant information such as the applicant's identity, employment details, income, and credit history.

Impact: Automation enhances processing speed, reduces human errors, and lowers operational costs for banks.

Fraud Detection

Real-time Monitoring: An AI system can analyze banking transactions in real time, identifying patterns suggestive of fraud. Machine learning models are trained on a database of fraudulent activity to detect characteristic signs of fraud.

Prevention and Protection: By utilizing these systems, banks can minimize or avoid substantial losses due to fraud and protect their customers, maintaining the reliability and credibility of banking institutions.

Credit and Lending

Credit Scoring Models

Advanced analytics enhance traditional credit scoring methods by incorporating additional data sources beyond conventional credit scores, such as rental payment history, utility expenses, and social media activity.

Improved risk assessment models provide a more accurate evaluation of a borrower's creditworthiness, reducing the risk of defaults and enabling banks to extend credit to a larger customer base.

Automated Decision Making

Efficiency in Lending: AI streamlines decision-making, particularly for loan application approvals. Analyzing applicant data can produce credit decisions in real time, significantly reducing the time from application to disbursal.

Accessibility: By speeding up decision-making and analyzing a broader range of data, AI enables banks to assess creditworthiness more accurately, allowing them to extend credit to individuals with lower traditional credit scores.

Looking Ahead: Challenges and Future Directions

As artificial intelligence continues to develop and penetrate various industries, we must address specific challenges to ensure its sustainability and safety. This section explores the costs and consequences of AI and the gap between its potential and user perception.

Costs and Environmental Impacts

Computational Costs

Resource Intensity: Advanced AI models, especially deep learning algorithms, require substantial computational power, relying on large data centers with high-performance, power-hungry servers.

Economic Costs: Supporting advanced computation incurs significant hardware and energy expenses, leading to high initial investments and ongoing costs for electricity and maintenance (Jones & Easterday, 2022).

Environmental Impact

Carbon Footprint: AI consumes a significant amount of energy, and most data centers use energy from fossil resources, which results in a considerable carbon footprint.

Sustainability Initiatives: There is a growing trend to make AI less energy-intensive by using new, efficient algorithms, employing green power, and increasing the efficiency of data centers.

Mitigation Strategies

Algorithmic Efficiency: Scientists are developing better algorithms that require less computational resources, which might lower environmental effects.

Renewable Energy: Raising the percentage of renewable energy sources used in data centers may help reduce the carbon emissions connected with AI training and functioning.

BRIDGING THE DIVIDE BETWEEN AI AND HUMAN COMPREHENSION

Complexity of AI Systems

"Black Box" Nature: Many AI systems, especially those that use deep learning technologies, are perceived by users as non-interpretable. This lack of transparency can undermine trust, particularly in critical areas like healthcare and self-driving cars.

Clarity in Decision-making: There is a growing demand for transparency in AI decision-making and a need for tools that empower users to understand how specific conclusions are reached (Asif, 2023).

Human-AI Collaboration

Complementary Strengths: Combining and leveraging the unique capabilities of AI and its individual users is the best strategy to achieve sustainable results and establish human trust. While AI excels at analyzing vast quantities of data, humans possess an unparalleled understanding of context, ethical nuances, and decision-making processes.

Interactive AI Systems: Implementing AI systems capable of explaining their decisions and incorporating a feedback loop can improve usability and boost efficiency.

Ethical and Social Considerations

Bias and Fairness: AI systems can perpetuate or exacerbate existing prejudices, leading to biased outcomes. Addressing this requires ongoing data remediation, continuous monitoring, and training AI models to prioritize fairness and equity (Asif, 2023).

Regulations and Guidelines: Comprehensive regulatory structures should oversee the development and application of AI to prevent the abuse of AI technologies.

Key Takeaways

The nature of deep learning technologies lends them to use in various applications. These promise to have a decisive impact on their industries and improve the quality of human life, provided that the technologies are properly trained and applied.

- **Imaging Diagnostics:** Convolutional neural networks are used in deep learning models to analyze medical images and diagnose diseases, tumors, and fractures quickly and with highly accurate results.

- **Personalized Treatment:** By utilizing patient data, AI can improve the effectiveness of suggested treatments based on specific patient health profiles while reducing possible side effects.

- **Molecular modeling** enhances drug identification by simulating molecular interactions and effectively modeling the results.

- **Clinical trials** use AI to optimize the selection of participants and improve the predictive analysis of their designs.

- **Remote monitoring** uses smart devices to track patients' health status and provide early treatment.

- **Virtual health assistants** provide initial analysis and medical consultation 24/7 by utilizing natural language processing functions.

- **Sensor fusion and real-time decision-making** describe how deep learning processes data from multiple sensors in real time and makes navigational decisions for self-driving cars.

- **Environmental understanding and object detection** explain how deep learning enables automobiles to understand their environment and safely maneuver the roads.

- **Facial recognition** technologies powered by AI work in applications for law enforcement and retail customer identification.

- **AI in finance** is transforming investment and asset

management strategies with trading algorithms and robo-advisory applications.

- **Predictive analytics in financial markets** use AI to predict future market trends and process complex financial analyses to make better investments.

- **AI's impact on traditional banking** covers a range of activities, from interacting with customers to improving back-end processes.

- **Data privacy concerns** can arise about data protection and the use of large patient datasets for deep learning.

- **Bias in medical AI** due to insufficient training data is a risk that can lead to healthcare disparities if it is not proactively addressed and remedied.

Looking to the future, the progress of deep learning across various domains presents both opportunities and risks. These technologies promise to dramatically reshape many aspects of human life and global economies. However, they also pose ethical dilemmas and underscore the need for consistent regulation. Specific steps are required to ensure that AI advancements align with human values. While this chapter underscores the profound effects of deep learning, it also supports cautious optimism in shaping its path forward. Adopting an inclusive approach that considers both the social and technical dimensions of deep learning is essential to maximize its positive influence on society.

SHARING YOUR LEARNING ABOUT AI: INSPIRING OTHERS TO EXPLORE NEW HORIZONS

"Artificial Intelligence is not just a tool; it's the new electricity, transforming industries and powering the future." — Andrew Ng

Dear Reader,

As we progress through this exploration of AI, it is clear that understanding this technology is crucial for navigating today's rapidly evolving world. My aim with this book is to provide you with practical knowledge that can unlock new opportunities and give you a competitive edge.

Your feedback is critical. By leaving a review, you contribute to a broader understanding of AI, helping others to discover essential insights that could benefit them. Your thoughts could guide someone just beginning their journey, giving them the confidence to explore AI's potential.

As someone who is passionate about AI and has extensive experience in IT, I hope this book will serve as a valuable resource for you. Your review is more than feedback; it is a way to support others in their learning and promote a community of innovation.

If this book has sparked new ideas or opened doors for you, please consider sharing your thoughts with others who are just starting to explore the fascinating world of AI technologies. Your contribution can help keep the momentum of learning and innovation going strong.

With sincere thanks,

Alex Quant

4

PROMPTING AND WORKING WITH CHATGPT AND OTHER AI

Mastering AI Communication

E ffective communication is vital to wielding the power of transformative technologies such as ChatGPT in the rapidly evolving AI landscape. This section delves into the nuances of interacting with AI, covering the structure of these interactions, strategies for formulating clear questions that are comprehensible to AI, and approaches to resolving any potential issues that may arise during engagement (Brockman, 2019).

Understanding the Anatomy of AI Communication

To facilitate effective communication with AI, it must receive proficient inputs, interpret them, and generate appropriate outputs in response to users. Let's examine the foundational elements essential for successful communication with AI.

Input Processing

The way that AI processes input determines the quality of its responses. Recognizing how systems handle this step can significantly enhance interaction quality.

Natural Language Understanding (NLU) is the basic process through which AI grasps human language. It involves analyzing text to determine its structure, keywords, and the user's intent, enabling AI to extract useful information from diverse linguistic data (Chacko, 2023).

Intent Recognition: AI uses NLU to perform intent analysis to determine whether the user is making an inquiry or a request, seeking help, or informing the system.

Information Extraction: In addition to intent, AI identifies information like dates, names, numbers, or any other details that it finds relevant in formulating an appropriate response.

Context Management

Especially in extended interactions, managing context is essential for maintaining a coherent and relevant dialogue.

Session-Based Context: In session-based interactions, knowledge is retained throughout the conversation. This continuity allows the AI to build on previous exchanges, ensuring coherence and making interactions smoother and more effective.

Ongoing Context Adaptation: Some of the most sophisticated AI systems can adapt to new contexts throughout an interaction and when they obtain new information. This adaptability is useful when historical data is required or when the topic of a conversation changes frequently. ChatGPT has recently developed a feature that enables its memory of content across chats. However, just like when you speak with a person, any discussion with AI should be framed in the current context, even if you ask it to retrieve content referred to in other chats.

Contextual Relevance: AI considers the context and decides on an adequate response. This might include appealing to previous parts of the conversation or inviting outside information that enriches the answer. Technologies like ChatGPT can search the Internet in real time for material to use in their responses.

Response Generation

Once the AI has processed the input and understood the context, it generates a response.

Natural Language Generation (NLG): This technology allows AI to form a sentence that is both syntactically correct and semantically relevant to the ongoing dialogue. NLG identifies suitable phrases and constructs sentences to create coherent and linguistically continuous text.

Content Tailoring: AI uses information about the user and the search query to provide the most relevant answers, recommend actions, or pose clarifying questions. It can also consider patterns and preferences it has gathered about user preferences to tailor its responses.

Dynamic Interaction: Superintelligent machines can adapt their responses based on user feedback. For example, an AI will correct its behavior in subsequent communications if a user marks a response as unhelpful.

By mastering input processing, context management, and response generation, AI systems can replicate human-like communication patterns that are natural and engaging for users. Understanding this structure of AI communication is crucial for developers and users of AI technologies alike.

Techniques for Crafting Effective AI Prompts

To successfully engage with an AI-powered system, it is necessary to know how to ask questions that will yield valuable and understandable responses. Here are some refined techniques that can significantly improve your communication with them.

Clarity and Precision in Language

Explicit Instructions: To reduce misunderstandings, avoid general expressions and communicate what you require from the AI as concretely as possible.

Specificity in Requests: Specify the exact question you are trying to answer. For instance, instead of asking an AI to write about history, you might request that it briefly describe the events of World War II (HarpOnLife, 2023).

Loquacity: AIs know a lot and can be "talkative", which can be extremely useful and informative. If the answers you get are too long for your tastes, ask the AI to respond briefly, in a few sentences. If needed, you can progressively add more specific questions moving forward.

Utilizing Question Types Strategically

Closed-ended questions are helpful when you are writing and require factual information or a straightforward answer. For example, "Is Paris the capital of France?" This type of question will return an answer quickly and without further elaboration.

Open-ended questions are suitable for creating debate or when more comprehensive information is needed. For instance, "What were the main causes of World War II?" This format of questions allows the AI to consider all the probabilities and provide detailed answers.

Contextualization

Background Information: Include relevant details directly in prompts, such as, "Given the recent developments in renewable energy technologies that we have been discussing, what are the potential benefits of adopting solar power for residential use?" Remember that AI can make mistakes; you should check important information. Decisions you make based on its responses are your responsibility.

Topic Switches: To avoid confusion and to ensure that it does not mistakenly relate new questions to the wrong context, it is best to alert the AI when changing the subject. For instance, "Moving on, how does quantum computing affect data protection?"

Use of Keywords and Terminology

Relevant Vocabulary: If possible, use keywords relevant to the topic of your inquiry. AI is an expert on everything, so this strategy enables it to quickly grasp what the user is asking and retrieve the right answers.

Technical Terms: While conversing about specific fields, use descriptive language to help the AI understand the depth of the conversation and respond accordingly.

OVERCOMING COMMON CHALLENGES IN COMMUNICATION WITH AI

Clarifying Ambiguities

To reduce ambiguity in communication and ensure that AI correctly interprets what you want, it may make sense to change how you phrase your questions or deliver information. Particular strategies can minimize misunderstandings and enhance the effectiveness of your interactions.

Double-Check Clarity: Before entering a prompt, read it to ensure no ambiguous terms may confuse the AI.

Reformulation: If an AI misunderstands you, try rephrasing your request and providing more details.

Progressive Detailing: If you are not yet ready to formulate your request precisely, you can start with a general question like "Can you tell me more about climate change?" and then follow up with more detailed questions about its effects, possible solutions and statistics (Shrimankar, 2024).

Specific Examples

Guide the AI's answers by using examples in the questions you pose and help it understand the context of your question.

Clarification through Scenarios: Providing hypothetical situations relevant to your question can produce more relevant answers. If you seek information about travel plans, consider phrasing your question concretely, e.g., "What kind of travel itinerary is suitable for a family with young children looking for a balance between adventure and relaxation?"

Combining Techniques for Clarity

Balancing Refinement and Examples: Elucidate specific topics using a combination of detailed inquiries and illustrative examples. For instance, if the AI misinterprets a question concerning software development methodologies, you could respond with, "Could you explain how Agile methodologies like Scrum and Kanban enhance productivity?"

Feedback Loops: To improve comprehension and the value of the responses you receive, use feedback loops and invite the AI to ask a follow-up question if it has not understood something or needs more information (Miller, 2024).

Following these strategies in communication with AI may require more effort, but it can reduce frustration from misunderstandings and ensure that the responses you receive are more relevant.

Resetting the Focus

If responses become off-track, resetting the focus to ensure continuity and relevance may be helpful. Here are some more detailed strategies and examples of how this works:

Clear Contextual Cues

Guiding Back on Track: When the AI veers off topic, issue direct prompts to refocus attention. For example, if it starts discussing broader environmental impacts instead of your interest, you could say, "Let's narrow the focus to the effect on farming practices."

Rephrasing for Emphasis: Reinforce focus by restating questions or statements. For instance, if the discussion strays from economic policies to other areas, you can ask, "Returning to fiscal policies, how do they impact small businesses?"

Summarize the Conversation

Recapitulation: Especially for certain complex topics like education or technical support, where extensive information may lead to confusion, it can be helpful to summarize content that you have discussed with the AI. For instance, "We have already discussed the impact of climate change on polar regions. Now, let's delve into its effects on tropical forests."

Checkpoints: Establish points in the conversation where you reflect on and confirm your understanding of an AI's responses before proceeding further. These periodic checks in training or consultation are important to guarantee an AI has grasped fundamental concepts before moving on to other levels (Dane, 2023).

These methods allow the conversation to continue down the right lane, guaranteeing meaningful conclusions and user interaction throughout.

Simplifying Complex Queries

Breaking down complex questions on difficult subjects or specific demands can improve interactions with AI. There are several strategies you can use.

Decompose Questions

Step-by-Step Inquiry: Divide a large question into smaller components. For instance, if you are looking for information on the effects of climate change on the worldwide economy, you might first ask, "What is climate change?", then, "How does climate change affect agricultural yields?", and finally, "What are the economic ramifications of changes in agricultural yields because of climate change?"

Modular Questions: It is better to use modular questions focusing on different aspects of a more general question when posing questions. Suppose your general question is: "How can we improve our customer service experience?" Using modular questions like the following will help AI to generate concrete measures:

1. "What are some effective strategies for reducing customer wait times?"
2. "How can we train our staff to handle customer complaints more efficiently?"
3. "What technologies can we implement to streamline our customer support processes?"

4. "What feedback mechanisms can we use to understand our customers' needs better?"
5. "How can we personalize our customer service interactions to enhance satisfaction?"

Sequential Querying

Building Block Approach: Approach each response as a puzzle piece toward understanding the next segment of a multifaceted question. Begin the conversation with the basic ideas and advance towards more complicated information as the conversation continues.

Layered Learning: Ask a sequence of increasingly specific questions based on the new knowledge that the AI gathers on a topic. An example of layered learning using a sequence of questions can look like this:

General Topic: "Sustainable Energy Solutions"

1. "What are the main types of sustainable energy sources?"
2. "How does solar energy compare to wind energy in terms of efficiency and cost?"
3. "What are the latest advancements in solar panel technology that improve efficiency?"

Contextual Build-up: Make sure that each question builds on the context created by the answers to the previous questions. This common thread helps synchronize communication between AI and users on a particular topic.

These strategies for refining complex questions can improve the quality of human-AI interaction, the quality of the answers you will receive, and your overall user experience.

Using Synonyms and Alternative Phrasings

Your choice of vocabulary and phrasing can powerfully impact the quality of your interaction with AI. The particular formulation of a message or a question can provoke a more or less relevant reply. This approach is primarily about awareness and may take some experimenting. You can consider several strategies to effectively employ synonyms and alternative phrasings.

Vocabulary Variation

Keyword Substitution: For example, if the question "What are the benefits of renewable energy?" does not yield an appropriate answer, you may ask, "What are the advantages of sustainable energy?". Replacing "renewable" with "sustainable" can yield a broader range of relevant responses, as the term "sustainable" might encompass additional aspects like long-term viability and environmental impact.

Phrasing Variation: Reformulating a question like "Can you explain how solar panels work?", to "How do solar panels generate electricity?" can produce a different response. Rephrasing to focus on electricity generation can lead to a more detailed and precise explanation, emphasizing the mechanism of action.

Explain Concepts

Contextual Definitions: If you use a specific or unusual term that could be ambivalent due to multiple meanings, adding a brief explanation will help AI respond relevantly. For example, "Can you explain the significance of *hygge* (a Danish concept that encompasses a feeling of coziness and comfortable conviviality with feelings of wellness and contentment)?".

Analogies and Metaphors: Explain ideas by relating them to familiar concepts. For instance, you may ask for an analogy for meditation and get a response like: "Meditation is like tuning a radio to find a clear signal amidst static noise. Just as a radio filters out interference to bring you crisp, clear music, meditation helps you quiet the mental chatter, allowing you to focus and connect with your inner peace and clarity."

Comprehensive Descriptions

Detailed Descriptions: When you request information about a particular process or idea, provide explanations rather than only referring to specific terms. For instance, instead of posing a question like, "How does a car engine work?", the question could have been phrased as, "Can you explain how fuel combustion generates power in an internal combustion engine, including how this power moves the car?". This detailed description provides context and focuses on the specific aspects of the engine's operation, making it easier for the AI to reply relevantly.

Context Inclusion: You can include a particular context in your questions to enhance comprehension. Instead of "How do plants grow?", you may ask, "What are the essential processes and environmental conditions that enable plants to grow, including photosynthesis, nutrient uptake, and the roles of sunlight and water?".

Using synonyms and other rephrasings can help an AI provide more detailed and targeted responses.

Regular Updates and Feedback

In the interest of continuously updating AI communication systems, it is crucial to use feedback regularly to maintain functionality and continually improve their ability to meet user needs.

Feedback Mechanisms

Direct User Feedback: Users should be able to give their feedback about each conversation with the AI in the form of a yes or no button or a star rating. This helps to fine-tune and train the AI to improve its delivery and match customer expectations.

Error Reporting: Users should be able to report specific issues or errors regarding responses. For instance, if an AI misunderstands a question or provides incorrect information, users can report such a response and briefly explain the problem they encountered.

Continuous Learning: Many kinds of AI continually adapt to new information and user input rather than needing to be retrained, making it easier for such models to change their response patterns in favor of new trends and user preferences.

Feedback Analysis: User feedback helps identify trends or problems we can rectify by modifying the training data or algorithms used in the AI system.

User Behavior Tracking: Questions that users pose to an AI and the answers they receive in return can be used to improve a model, make it more appealing, and proactively target areas that it finds confusing.

Benefits of Regular Updates and Feedback

Enhanced Accuracy: New data inflows and user feedback provided through the interface increase the reliability of AI systems and minimize the likelihood of errors.

Personalized Interactions: Self-learning systems that receive feedback regularly can offer users a more relevant and engaging experience.

Increased User Satisfaction: AIs updated frequently based on user feedback are usually rated higher because they are more responsive and accurate.

Incorporating effective feedback mechanisms and continual learning into AI enables systems to remain relevant and improve communication performance. Such continuous evolution is necessary for maintaining user interest and confidence in AI solutions.

Adjusting Expectations

Setting appropriate expectations for AI is an important first step in ensuring successful collaboration. We can apply several strategies to manage and adapt our expectations to suit AI's capabilities and limitations.

Understand AI Capabilities

System Strengths: It is important to be aware of what a particular AI is best suited to do, whether that is handling big data, answering customer inquiries, conversing naturally, or creating unique content. Every AI system contains parameters to improve performance on concrete topics and tasks.

Recognize limitations: It is useful to be aware of the restrictions attributed to the AI that you are employing. For instance, some AIs could struggle with reasoning in context or comprehending humor like sarcasm or idioms.

Technical Boundaries: We should be mindful of an AI's limitations for long-term use. For example, we should understand that data privacy issues may exist when handling sensitive information like personal health records. It is also important to recognize that certain kinds of AI can take several seconds to generate a response, which could impact real-time applications. Depending on its specified purpose and training, an AI may need help to accurately analyze complex or specialized data, such as legal documents or scientific research papers.

Advantages of Proper Expectation Management

Enhanced Collaboration: When users know what AI can or cannot do, they can better incorporate it into their respective work environments, thereby optimizing productivity and the originality of work.

Reduced Frustration: Setting realistic expectations and patiently adapting your use of AI can improve the value of the tool and your overall user experience.

Improved Outcomes: People who align their interactions and expectations with an AI's advantages and disadvantages enhance their results because they know more about the AI's performance and can adjust to and optimize using it to get better responses.

You can make the most of AI systems in numerous activities and situations by maintaining realistic expectations and being ready to change your behavior to improve results.

Practical Applications of AI Technologies

Enhancing Productivity through Organization

Improving productivity with AI involves leveraging various tools and applications to optimize daily tasks and administrative functions. AI systems can impact productivity in many different areas.

Task Management

Integration with Digital Assistants: Advanced virtual assistants such as Cortana by Microsoft, Google Assistant, and Siri by Apple can schedule personal and business calendars. They can arrange appointments, provide alerts for tasks we need to accomplish, and even recommend good times to do something in light of an individual's routine and previous commitments (Kagan, 2024).

Predictive Scheduling: Some AI systems use historical calendar data to forecast working availability and provide the best times for scheduling meetings and tasks, avoiding multiple scheduling clashes and boosting efficiency.

Email Sorting and Response

Advanced Filtering and Prioritization: AI algorithms sort incoming emails based on the sender's content and level of importance, thus indicating which emails are most urgent and enabling professionals to prioritize crucial communication.

Automation: AI can automate appointments by independently managing scheduling, rescheduling, providing necessary information, and sending follow-up details, thus saving time and increasing efficiency without human intervention (Somers, 2023).

Document Organization

Intelligent Search and Retrieval: AI systems incorporate natural language processing to enhance search capabilities and determine the context of documents. This way, instead of trying to recall a file's complete name, a user can search for it by using keywords or content they remember.

Automated Filing: Some programs are designed to categorize documents and assign them to folders based on their content, keywords, or projects. This can be particularly helpful in legal, academic, and large corporate environments, where the efficient management of documents is paramount.

Version Control: AI can keep track of versions, indicate merges, and preserve the document version history, which is necessary when working with various documents simultaneously.

These functionalities support individual activities, collective work, and project management, where everybody gets updated information in the most structured form. AI helps free up the mental capacity of individual users and even entire teams and directs it towards complex problem-solving and innovative work, which in turn enhances productivity.

Creative Writing and Content Generation

We have already seen how AI is expanding the frontiers of content creation in education, but it continues beyond there. It is changing how content is written across all platforms, simultaneously speeding its production and enhancing creativity.

Story Plot Generation

Automated Storytelling: In its various iterations, ChatGPT uses literature and cinematic productions to build engaging storylines and character arcs. Writers can start by entering simple plot points or character attributes, and the AI can create entire plots complete with themes appropriate for specific genres.

Dialogue Enhancement: AI can recommend dialogue options that resonate with script tones and character voices, which can be especially helpful in multifaceted plots where it is essential to keep track of character voice patterns (Gunter, 2023).

Marketing Content Creation

SEO-Optimized Content: Software-aided content-generation tools study trends and keywords to produce optimized content that ranks well on search engine results pages (SERP). This pertains to creating meta tags, headings, and content that adhere to current best practices.

Automated Ad Copy: AI technologies can create multiple versions of ad text to test which ones are most effective, giving marketers a systematic tool to optimize their advertising efforts.

Content Scaling: AI assists businesses in amplifying content production at scale without necessarily diminishing the quality of the work produced. For instance, a single base article can be rewritten differently to target specific customer segments or regions so that content can reach a large audience.

Personalized Content

Dynamic Content Customization: Automated technologies use information and interactions to create responsive content on websites and other media. That can be as simple as changing the language according to a customer's location or as advanced as adjusting the structure of a site and the product suggestions it makes based on a user's past activity.

Interactive Media: In the context of video games and other interactive media, AI creates dialogues or quests that adapt and change based on a player's responses or actions. This increases user interaction and creates a wider range of opportunities for non-linear narratives.

Email Marketing Automation: AI enhances the effectiveness of email marketing by sending messages that reflect a user's product engagement. It can change the timing, content, and promotional offers to customers to improve sales conversion rates.

When implemented in creative tasks, artificial intelligence can use big data to create something unique to meet people's needs and expectations.

Learning and Tutoring

AI is revolutionizing education by delivering personalized content and adaptive tutoring.

Personalized Learning Experiences

Adaptive Learning Algorithms: AI can use students' performance data to determine the subjective difficulty level of the tasks given to them, their rate of learning, and the educational materials that should be used. This flexibility can help satisfy the needs of individual students and may lead to enhanced participation and achievement.

Curriculum Customization: Based on students' particular strengths, weaknesses, and preferences, AI can track and develop tailored learning plans and programs for them. This is especially true in areas of study such as mathematics and the sciences, where building on fundamentals is necessary to progress to the next level (Das, 2024).

Automated Tutoring Systems

24/7 Availability: AI tutors are available around the clock, offering assistance outside school hours. Students who need late-night assistance completing their assignments or preparing for exams can benefit from this.

Instant Feedback and Assessments: AI platforms can give students instant guidance about their work and help them to make corrections immediately, which is critical for enhancing learning and continuous improvement.

Intervention and Support: AI is able to pinpoint when a student has difficulty understanding learning material and apply targeted measures before they fall behind in the subject (Willmore, 2023).

Language Learning

Interactive Language Practices: The Duolingo language learning application uses artificial intelligence to test a user's abilities and assign adequate practice exercises. These include grammar checks, vocabulary exercises, and role-play.

Pronunciation and Listening Skills: AI-based speech recognition and processing technologies assist language learners in enhancing their pronunciation and listening abilities through feedback and correction.

Cultural Context Integration: Language teaching capabilities are enhanced by incorporating the cultural context features essential for achieving true mastery of a language into lessons (Chen, 2023).

These innovative applications of AI in education are changing conventional learning paradigms to make education more approachable, individualized, and streamlined. Teachers can use them to address a wide range of student needs and make the learning process more inclusive.

Key Takeaways

Effective communication with an intelligent assistant relies on well-formulated questions and clear instructions. We should know how natural language processing works and use it to our advantage. Addressing questions in context helps to ensure that AI can optimally understand requests and generate high-quality, coherent responses. Knowing how it processes inputs can empower us to use it effectively for various tasks, including content generation and improving productivity and learning processes.

- **Creating Effective Prompts:** Best practices for AI prompts include using qualifiers in your wording, choosing suitable types of questions, briefly referencing the context, and layering questions to enhance the depth and detail of the answers.

- **Overcoming AI Communication Challenges:** Remove ambiguity by refining terms, restoring context where necessary, breaking down questions, and rephrasing.

- **Practical Applications of AI:** Optimize work processes by automating organizational tasks and promoting learning with tailored educational applications and intelligent tutors.

- **Enhancing Productivity through Organization:** AI can help schedule, filter, and organize emails and documents to improve work efficiency and productivity.

- **Creative Writing and Content Generation:** AI improves creativity by suggesting ideas, writing texts, customizing content for target audiences, and interacting with users.

- **Enriching and Tailoring Educational Experiences:** AI offers learning flexibility and constant guidance to meet individual educational needs and preferences.

Using AI-powered tools provides numerous opportunities to optimize organization, spark creativity, and achieve personal fulfillment. The social applications of algorithmic technologies are increasingly important, particularly in education and workplace efficiency. This chapter aims to raise awareness about the factors that can enhance communication and collaboration with AI and provide examples of how this technology has transformed essential aspects of our daily lives.

5

AI IN THE DIFFERENT SECTORS: WHAT YOU NEED TO KNOW

AI in the Workplace

As workplaces evolve, AI is becoming a crucial factor in business operations. It enhances efficiency by automating tedious and repetitive tasks and improves effectiveness in complex decision-making. As such, it is transforming tasks and redefining roles and learning environments in the workplace. This section explores various aspects of AI in the workplace, including process automation, project management, decision support, work redesign, and learning and development.

Automating Routine

The use of AI for automating mundane tasks is transforming conventional work paradigms and enhancing organizational processes.

Process Automation

Streamlined Operations: AI technologies perform valuable but routine work, including filtering large quantities of information while handling transactions and records with minimal error rates. For example, they can be applied in financial services to automate and speed up the loan or claims process, which used to take days or weeks.

Intelligent Algorithms: Computerized algorithms can perform extensive data analysis quickly and make decisions based on specific parameters. In customer service, AI chatbots reply to customer inquiries and address their needs around the clock.

Error Reduction: AI helps to minimize human errors in routine processes, which is especially useful in disciplines like accounting and data input, where accuracy is paramount (Radley, 2024).

Robotic Process Automation (RPA)

Human-Like Interaction: RPA tools mimic user interactions with digital systems and can execute activities like form-filling, data extraction, and file transfers. For instance, a company might use an RPA tool to process invoices received via email. RPA can significantly increase the pace of activities like training new employees or updating customer information.

Scalability: Unlike human employees, RPA systems can scale up or down based on demand. They can handle higher loads when businesses receive a flood of requests.

Integration: RPA connects legacy and new systems through interactions that do not entail robust API connections, making it a valuable tool in organizations with outdated operating systems (How can you use AI to automate WFM tasks? n.d.).

Enhanced Efficiency

Time Management: Automation enables organizations to deliver services faster, increase customer satisfaction, and gain a competitive edge. For example, in e-commerce, AI systems help determine the most efficient routes to minimize delivery time.

Cost Reduction: Automating repetitive tasks decreases the demand for extra staffing, thus saving overhead costs. AI systems do not require rest; they can work 24/7 and provide consistent output.

Focus on High-Value Work: By automating repetitive work, AI frees human resources for more complicated and creative assignments. As such, it can promote higher job satisfaction and foster innovation in organizations (The Role of AI in Workforce Management Processes, n.d.).

Strategic Implementation

Custom AI Solutions: Organizations increasingly seek customized and specific AI solutions for their operations. These solutions can automate particular tasks with ready-made applications. For instance, an aerospace components manufacturer could use a tailored AI application to manage quality control for its highly specialized, complex parts.

Continuous Improvement: Many AI systems contain machine learning functions that enable them to adapt and learn from new inputs and experiences, gradually increasing productivity.

Artificial intelligence enables the automation of many tasks, leading to increased efficiency and productivity. This transformation alters how tasks are executed and reshapes roles, making workplaces more dynamic and focused on innovation and strategic initiatives.

AI Tools for Project Management and Decision Support

Organizations increasingly use AI solutions to improve project management and decision-making processes.

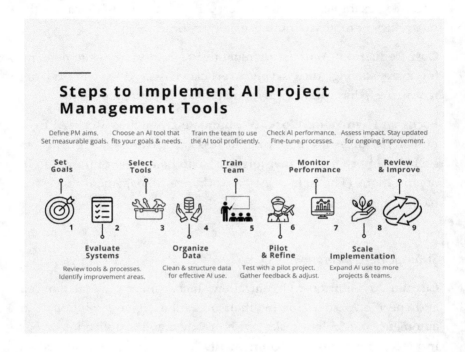

Fig. 4: Steps to Implement AI Project Management Tools

AI in Project Management

Predictive Planning: Based on data about similar projects, AI can predict when a project is likely to be completed and its success rate, helping project managers realistically assess time frames and estimates.

Resource Optimization: AI tools can categorize work patterns and resource utilization to recommend appropriate allocations. They can predict periods of heavy traffic and advise about the most effective use of human and other resources.

Task Automation: AI can support simple activities like updating a project's status, providing reminders, or creating reports, thereby relieving project managers of these administrative tasks and allowing them to concentrate on strategic project management at the macro level.

Real-Time Problem Solving: AI systems can track a project's progress in real time and inform managers about delays or cost over-runs. Advanced tools can also suggest corrective measures based on an organization's past performance.

Decision Support Systems (DSS)

Data-Driven Insights: Analytical DSS tools use AI to process extensive information that assists in optimal decision-making. They can help forecast market trends, consumer behavior, and other economic factors influencing business decisions.

Scenario Simulation: AI can perform modeling based on various business parameters, making it easier for executives to minimize uncertainty and assess scenario implications for better decision-making.

Enhanced Visualization: Data collected through AI can be transformed into easily understandable graphical forms, enabling executives to gain insights and make informed decisions faster (Kayser, 2024).

Risk Assessment and Management

Predictive Risk Analysis and Mitigation Strategies: AI is able to use data from previous projects to predict possible risks, such as project delays or cost increases, and it can also recommend the best ways to address them. This proactive approach addresses emerging challenges before they get out of hand.

Continuous Risk Monitoring: By continually monitoring project parameters, AI allows project managers to change risk assessments quickly without compromising a process's flexibility and speed.

Integration and Collaboration

Software designed for assimilation and interoperability is essential in today's workplace, where seamless technology integration and collaboration are vital components of productivity and efficiency.

Cross-Platform Integration

Seamless Connectivity: AI tools are compatible with almost any software platform, including enterprise resource planning, supply chain management, and CRM, ensuring that they do not interrupt programs that are already in use.

Data Synchronization: Integration enables real-time data feeds across various media. Changes made in one application are immediately propagated to the others, contributing to better data consistency and minimizing data replication.

Plug-and-Play Features: Some existing AI solutions are designed to integrate seamlessly into existing systems by interfacing via APIs. Businesses can incorporate technologies into their operations without significant modifications to their IT systems.

Collaborative Features

Team Member Recommendations: Based on past results and employee scores, AI can suggest which employees should be involved in which projects, thus improving the composition of a team and project results.

Automated Coordination: AI can easily plan meetings, organize tasks according to an employee's work capacity and experience, and send alerts for deadlines, streamlining project operations without the need for constant senior supervision.

Enhanced Communication: By integrating AI into the communication interface, messages can be analyzed based on urgency to prioritize critical communications. It can also recommend appropriate replies to standard and otherwise time-consuming emails.

Conflict Resolution: Since AI can identify conflicts within a team and analyze communications and sentiment, it can alert a manager when an intervention is needed or even provide possible solutions to avoid conflict escalation.

Real-time Collaboration

Document Collaboration: Algorithmic tools facilitate the simultaneous, real-time editing of documents and projects, allowing teams to work together more effectively. These tools monitor changes, suggest improvements, and predict potential problem areas based on previous collaborations, enhancing team efficiency.

Virtual Workspaces: Artificial intelligence creates dynamic virtual work environments that adapt their layout based on workflow, project priorities, team interactions, and available resources. This flexibility enhances collaboration and efficiency by aligning the workspace with the team's evolving needs.

Predictive Collaboration

Anticipating Needs: Thanks to its predictive capacity, AI can foresee what resources a team may require to execute a project, thereby avoiding interruptions and improving the efficiency of the outcome.

Feedback and Learning: Algorithmic systems capture team responses and project outcomes to refine their suggestions and behaviors. This continuous feedback loop enhances the system's ability to support team collaboration and improve efficiency over time.

Introducing intelligent applications allows organizations to benefit from improved connectivity, automation, and advanced predictive modeling for project management. These innovations boost productivity, enhance team skills, and shift human focus from routine tasks to strategic thinking and advanced problem-solving.

Strategic Decision Support

We have touched on how AI enhances decision-making by offering key insights and forecasts. A number of its capabilities ideally support strategic planning and custom decision model development.

Long-term Planning

Trend Analysis and Forecasting: Algorithms can analyze large amounts of historical data and find patterns that might be hard for human analysts to uncover. This capability enables organizations to predict future market trends and changes in consumer behavior and technology, which is helpful for long-term planning.

Scenario Planning: Based on different hypothetical strategic decisions, AI is used to model multiple future scenarios and prepare organizations to prepare in advance.

Resource Allocation: Since intelligent systems can estimate when resources will be needed in the future, they are able to recommend the most suitable way to allocate them between multiple projects or departments to achieve a better ROI.

Custom Decision Models

Tailored Algorithms: Enterprises can create AI models tailored to the strategy of the organization. These models incorporate the specifics of the particular business environment, its competitors, and the internal organizational culture.

Integration with Business Intelligence: Incorporating AI models into Business Intelligence systems gives organizations a broader perspective of their operational environment.

Dynamic Adaptation: Self-learning models can evolve with time and exposure to new data and environments. This dynamic adaptation guarantees that decision support is current and accurate as a business changes.

Decision Optimization

Cost-Benefit Analysis: Current models can perform detailed cost-benefit analyses far more efficiently than traditional methods. By evaluating decision risks and opportunities, AI enables more rational and cost-effective choices.

Multi-Criteria Decision Analysis (MCDA): Using smart technologies to integrate multiple factors into the decision-making process simultaneously, comparing costs, risks, and benefits, determines optimal courses of action.

Supporting Executive Decision-Making

Executive Dashboards: Executive summary reports developed by sophisticated AI offer boardroom-level data and KPIs relevant to particular executives' concerns. These dashboards may help decode complex data and enable faster and better decision-making.

Real-time Insights: Thanks to real-time processing, AI allows us to present up-to-date information in an ever-changing business environment.

Through the application of AI for predictive analysis and strategic decision-making, organizations can gain a competitive advantage and shape their future course of action. This proactive strategy is highly relevant in today's constantly evolving and sometimes volatile business settings.

The Evolution of Work

Integrating intelligent systems into business processes marks an unprecedented evolution that will impact the work environment in all aspects, roles, and skills. Next, we will examine how AI is reshaping the modern workplace.

Job Redefinition

Task Automation and Role Adaptation: AI's ability to process repetitive tasks will allow employees to work on more critical assignments. As a result, job requirements will evolve to include more creative, analytical, and interpersonal activities that are beyond AI's capabilities.

Creation of New Roles: As intelligent applications are increasingly integrated into organizational processes, new positions that deal with the design, support, and governance of AI are being created. These include roles such as AI Trainers, Ethics Officers, and Integration Specialists (Ortiz, 2024a).

New Skill Requirements

Technical Proficiency: Even if they are not working in a technology-based position, it is becoming imperative for employees to have at least a basic understanding of artificial intelligence and machine learning, since they are increasingly required to use these tools to perform their tasks.

Emphasis on Soft Skills: Since AI is likely to take on more computational and repetitive tasks, skills such as creativity, emotional intelligence, and communication remain valuable. Within an organization adopting AI systems, they are more essential than ever in team management, decision-making, and customer service.

Ethical and Responsible Use of AI: As AI is integrated into many business contexts, it is important for employees to know about its ethical concerns, including privacy, fairness, and responsibility. Training in the ethical use of AI guarantees that employees are equipped to make sound choices that serve the best interests of the company and its customers (Wiggers, 2024).

Collaborative Human-AI Work Environments

Cooperative Systems: Intelligent solutions are being developed to enhance human capabilities. While AI processes algorithmic aspects of jobs, people handle other aspects that require emotional and analytical intelligence. This synergy can improve efficiency and creativity in different fields of business.

Human-in-the-loop Systems: In cases that require delicate handling, AI should be directly supervised by human beings who can override its decisions and take the final verdict. This structure minimizes errors and biases.

Training for Effective Collaboration: Training is emerging in the workplace on interacting with AI, including interpreting its results and managing algorithmic tools. Employees need this knowledge to

ensure they can effectively use the technologies without being domi-
nated by them.

Impact on Organizational Structure

Flatter Organizations: As it automates many monitoring and data
processing responsibilities, AI can lead to the reduction of middle
management.

Dynamic Teams: Visionary approaches enable using artificial intelligence to form project teams for specific tasks based on their skill sets
and availability to improve operations and project performance.

Cultural Shifts

Adaptation to Continuous Learning: Given the rapid advancements
in technology, the workforce must embrace continuous learning and
flexibility. Organizations will need to invest more in training and
development activities to ensure employees maintain relevant skills
and effectively utilize new AI applications, enhancing overall
competitiveness.

Work-Life Integration: Arrangements like teleworking and on-
demand working, which can improve work-life balance and entail
changes in work and employment relationships, can be enabled
by AI.

Integrating intelligent systems into the workplace fundamentally
affects the nature of jobs and skill requirements as well as the culture
and structure of organizations. This level of continuous change
requires flexibility and the ability of training and development divisions to proactively prepare staff and organizations for the new world
of work.

Learning and Development in the Era of AI

The advent of AI represents a paradigm shift in training and development that will empower employees through increasingly innovative programs.

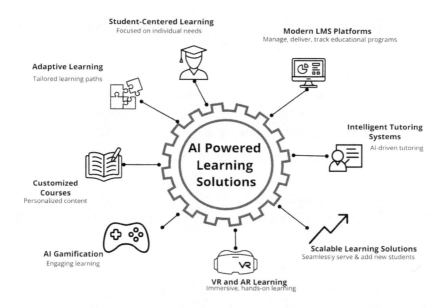

Fig. 5: AI for Learning and Development

Personalized Learning

Tailored Educational Experiences: Similar to its potential in formal education, AI can also be used for training in the workplace. Taking account of an employee's learning behavior, performance, and preferences, educational materials can be personalized to fit their needs. This customization enables learners to work through material at their own pace while focusing on elements that need attention, enhancing learning efficiency and effectiveness.

Adaptive Learning Paths: Autonomous systems dynamically adapt the difficulty and structure of training materials based on performance. They can take account of an employee's strengths and weaknesses and adjust training plans accordingly (How can you use AI to personalize employee development? n.d.).

Continuous Professional Development

Skill Gap Analysis: In HR, AI tools can evaluate an employee's fit for a current or future role and define which abilities need to be developed. This makes training more focused on meeting the needs of both the employees and the organization they work for.

Automated Recommendations for Training: AI considers constant performance reviews and shifts in business requirements when providing personalized courses and learning modules to learners. This proactive approach ensures that skills remain relevant even in an environment of constant change.

Microlearning and Just-in-Time Learning: Artificial intelligence endorses microlearning approaches that provide employees with short, targeted learning sessions that can be accessed on-demand. Since the information they convey is immediately usable and they lend themselves to seamless integration into work processes, a high rate of employee attention and retention accompanies them.

Cultural Adaptation

Fostering a Learning Culture: Leaders can harness AI to develop a culture that promotes continuous learning and skill development in the workforce. Smart technologies can positively influence learners' engagement by demonstrating the relevance of continuous education to their careers and providing them with optimal efficiency.

Change Management: AI forecasts the consequences of organizational changes, such as implementing new tools or procedures, making it easier for organizations to plan and train their people, mitigate potential resistance, and improve the implementation of change (Hemmler & Ifenthaler, 2022).

Integration with Human Resource Processes

Enhanced Onboarding: AI makes the training process for new employees easier by customizing programs based on their past experiences and the requirements of new positions in an organization, which saves time in getting new employees up to speed with other team members.

Career Development Planning: When intelligent technologies match individual goals with organizational requirements, career development planning becomes more efficient, and employees can gain new opportunities (Kobylinska, 2023).

Impact on Learning Formats

Virtual and Augmented Reality Training (VR and AR): VR and AR can be combined with AI to enhance learning experiences through simulation, which is especially relevant in technical areas such as machinery operations or procedural training. AI-enabled applications adjust scenarios and difficulty levels according to a learner's pace and efficiency.

Interactive Simulations: Virtual reality-based training allows employees to practice skills and decision-making without negative consequences. These simulations change dynamically in response to learners' decisions. They offer customized feedback and diverse scenarios to improve the effectiveness of learning processes.

Ethical Considerations and Accessibility

Equal Opportunity: Since AI systems learn from the data provided to them, using fair and non-prejudiced algorithms in the delivery and assessment of training is critical to ensuring equal chances in the workplace.

Accessibility Enhancements: AI improves access to learning in the workplace for people with disabilities. For instance, it transcribes video and audio recordings to assist visually impaired persons (Bird, 2023).

The profound shift represented by AI in learning and development will fundamentally reshape how training and skill enhancement are viewed and implemented. By providing essential tools and opportunities for advancement, the technology is opening up new career paths for employees. The strategic application of AI in human resources enables organizations to stay relevant and effectively meet the ever-evolving challenges of the global market.

AI in Information Technology

AI is a disruptive element in Information Technology (IT), influencing how this sector functions, processes data, and enables business operations. It is impactful because it can run intricate procedures, assess massive quantities of information, and promptly make rational choices based on these analyses. These capabilities are essential for increasing efficiency, minimizing the impact of human mistakes, and speeding up operations within IT environments.

Why AI Impacts IT

AI has greatly influenced Information Technology broadly and deeply, changing the nature of computing departments and bringing compelling advancements to the field.

Automation of Routine Tasks

AI's ability to automate, which we have seen prominently in other spheres, also marks its most significant impact on IT.

Task Automation: Intelligent systems can handle multiple IT tasks, such as installing and updating commonplace programs and configuring and optimizing networks. This automation is more than mere pattern replication; it includes the ability to adapt based on current data.

Efficiency and Productivity: As in other spheres, AI can free up the schedules of IT specialists to focus on more meaningful projects that bring higher value to a company. It is especially notable for eliminating overhead costs and mistakes made by people in complicated supply chain processes.

Enhanced Security Measures

AI enhances IT security in several critical ways.

Threat Detection and Response: Intelligent systems look for patterns in network traffic that deviate from the norm and could indicate a security threat. These systems can effectively recognize malware. They use machine learning algorithms which are trained with historical data and improve their ability to detect threats over time.

Real-time Security: Unlike conventional security systems, which are periodically updated, AI-based systems work in real time. They can swiftly react to developing threats, thus considerably lowering potential damage.

Automated Security Protocols: AI can launch or recommend defensive measures without involving security teams, saving time and effort and reducing damage thanks to its rapid response.

Predictive Analytics and Optimized Resource Management

The predictive capabilities of AI are redefining how IT departments handle maintenance and planning:

Anticipating System Failures: Using pattern recognition, machine learning can tell when hardware will likely fail or software may crash. This predictive maintenance makes IT systems more reliable and reduces costs associated with unexpected breakdowns (Importance of artificial intelligence (AI) in information technology, 2020).

Resource Allocation: AI-based analytics and monitoring can predict future IT demands, such as support needs, bandwidth, and processing power, and automatically distribute them according to requirements, making it easier to prevent performance bottlenecks and ensure the efficient utilization of all computational resources.

Energy Efficiency: As AI effectively manages IT assets in data centers, energy use is also effectively controlled – a vital consideration in controlling costs and promoting environmental conservation.

Support for Decision Making

As in other sectors, AI is also reshaping the way decisions are made in IT, making them more data-driven:

Insightful Analytics: Intelligent tools work with large volumes of information and can identify relationships that would remain unnoticed by a human analyst, offering IT managers a detailed understanding of system status, user activities, and possible risks.

Strategic Planning: AI can help IT departments accurately predict future needs, budget for software and equipment, and mitigate potential threats.

Enhanced User Experience via Custom Software Solutions

Artificial intelligence significantly improves the user experience provided by IT services by paving the way for more sophisticated, context-aware software solutions:

Predictive User Interfaces and Personalization: So-called predictive or smart user interfaces can adapt to a user's work habits and anticipate their next move. AI can then tailor real-time adjustments at the user level, customizing the desktop environment and recommending content based on a user's role and interactions within an IT environment, thus greatly improving workflows and user satisfaction.

User Behavior Analysis: Since AI systems adapt applications to how users utilize them, they note inefficient designs and can change them dynamically.

Automated Support for End-Users: Chatbots and virtual assistants can promptly respond to multiple IT requests from end-users, relieving the human support team and improving user satisfaction.

Strategic Impact and Competitive Advantage

AI influences the strategic role of IT within organizations, significantly enhancing the efficiency of computing departments.

State-of-the-art Services: AI supports the rapid development and implementation of new IT solutions. It enables IT departments to explore new technologies and adopt them into organizational structures more quickly and proficiently, helping organizations to excel and stand out in the market.

Business Alignment: Using AI, IT operations and decision-making can better align with business goals. Forecasting analysis and intelligent automation help computing facilities develop and meet business requirements, thus contributing to growth and increased competitiveness.

Thanks to its ability to handle routine tasks, enhance security, and provide predictive analytics, AI optimizes IT department operations and elevates their significance within organizations. As AI advances, its adoption in computing practices not only aligns organizations with current technological advancements but also enhances the role of information technology in achieving business objectives. Intelligent systems are essential for maintaining and improving current processes, discovering new opportunities, and achieving maximum efficiency.

How IT Development is Influenced by AI

AI is becoming a game changer for IT development and is integral to many aspects of software production and support. It is proving to be a potent driver for innovation, impacting every stage of the software development lifecycle and creating a fundamental shift in how solutions are engineered.

Accelerated Software Development: AI can automate many aspects of software development, including code generation and quality assurance, thereby speeding up the software development life cycle (SDLC) and allowing IT professionals to focus on other critical issues.

Enhanced Problem Solving: AI identifies and solves IT issues faster than conventional processes, increasing efficiency for IT personnel, who are freer to address other tasks.

Requirement Gathering: Natural language processing is used to translate and analyze software requirement documents and recommend features that ideally address business needs.

Design Phase: AI design tools draw on best practices and previous experiences from similar projects to create a blueprint for new software.

Maintenance: Machine learning consolidates learnings about software performance and user satisfaction to target constant improvement and enhancement.

Optimized Software Testing

The efficiency and effectiveness of software testing processes are significantly enhanced by using AI.

Automated Test Case Generation: Algorithms can automatically develop test cases from the codebase and its usage patterns, eliminating the need for manual effort to test all the functionalities.

Prototype Testing: New technologies can be tried and realistically assessed in simulated model environments before large-scale implementation. AI automates integration and deployment, enabling more frequent software releases with fewer mistakes.

Enhancing Code Quality and Maintenance

AI tools are also revolutionizing code development and maintenance.

Predictive Bug Tracking: Models can identify which areas of an application are likely to encounter problems in future updates by analyzing the history of a project's code modifications and bug reports. This proactive approach helps developers anticipate and address issues before they become critical.

Smart Test Execution: Intelligent agents can sort test cases based on the order of code modifications and previous test outcomes, targeting areas with potentially faulty code. This approach speeds up the testing process and reduces costs by using resources more efficiently, ensuring that testing efforts are focused where needed most.

Automated Code Review: Tools powered by artificial intelligence exist that scan through code to identify flaws and propose enhancements.

Legacy Code Modernization: The legacy codebase can be reviewed and refactored with AI without constant manual monitoring.

AI is transforming IT development, presenting new growth opportunities, and creating increasingly intelligent, adaptive, and user-oriented software systems, ensuring it remains an integral part of computer development.

AI in the Creative Industries

In the creative industries, artificial Intelligence is setting new standards as it proves its value as a generative enhancement in art, writing, music, film, and gaming.

Fig. 6: AI Can Be Used to Co-Create a Range of Artistic Styles

AI in Art and Design

One of the biggest questions surrounding AI is its potential for creativity. Traditionally, art has been considered a uniquely human endeavor, deeply influenced by emotion, spirituality, and subjective expression. Nonetheless, AI can generate work that embodies aspects of creativity through its algorithms and the data it processes. By analyzing vast datasets of diverse artistic influences, styles, and techniques throughout history, AI can produce works of art as original and provocative as the user's imagination. This capability challenges the conventional belief that creativity is exclusive to humans (DeSoto, 2024; Wang et al., 2024).

Art created by prompting AI raises questions about the nature of artistry and genius and the possibility that AI-human co-creation can expand the idea of art and creativity.

How AI is Impacting Art and Design

Machine learning approaches are now widely used in art and design to increase productivity and creativity. Tools equipped with AI capabilities can analyze and replicate complex patterns and styles, assisting artists and designers in various ways.

- **Rapid Prototyping:** AI streamlines the last phase of a designer's work, making it faster to prototype and adjust their designs.
- **Pattern Generation:** The technology can quickly create detailed and complex patterns, which is useful in the textile industry and in graphical and architectural design.
- **Exploration of New Forms:** Artificial intelligence expands the possibilities for art and design by allowing users to experiment with hybrid arts and computational creativity.

Applications such as Adobe Sensei enhance productivity through features like auto-cropping photos and enhancing detailed vector graphics, allowing designers more time to brainstorm in their projects.

Using AI to Elevate Art

The application of AI in art is about more than just automating work. It can also increase interaction and engagement. More and more, artists use it to create installations that adapt and morph in response to various stimuli.

Interactive Installations: Some artists use AI to create installations that can change in response to viewers and the environment. For instance, AI can modify the graphics and audio in an artwork depending on an audience's movements, the sound of voices, or even facial expressions (Williams, 2023).

Dynamic Artworks: Some pieces are designed to develop over time by incorporating AI-powered elements in the compositions. These can include features that reflect changes in social media trends, news feeds, or weather applications, producing a dynamic art form that evolves and can comment on contemporary worldwide developments (Shahzad, 2023).

Artificial intelligence has emerged as a technology that builds on the foundations of classical artists and empowers ordinary individuals to create without traditional training. This innovation has given rise to a new form of art that merges human creativity with AI, pushing the boundaries of what we understand as artistry.

Fig. 7: AI Can Help Generate Photorealistic Images as Well as Fantasy Artworks

AI in Writing and Content Creation

Thanks to automation, AI is having a major impact on writing. The Generative Pre-trained Transformers (GPT) developed by OpenAI exemplify how AI can contribute to various forms of composition.

- **Content Generation:** Application intelligence tools have enabled content writers to produce everything from poems to complex analytical articles.

- **Idea Generation:** AI can overcome creative stagnation in many ways, whether it inspires new ideas, fresh perspectives, or develops styles created by other artists.
- **Drafting Assistance:** By incorporating details or amplifying critical points, AI can complete drafts, allowing authors to focus more on narrative development (The future of creative writing with AI technology, 2024).

Earlier, we considered the impact that artificial intelligence is having on personalized content and marketing copy. It is particularly effective for creating advertising material, which involves generating high volumes of content targeted to appeal to particular audiences. While these two areas of application are evident, AI is also increasingly being used successfully in other creative realms.

Algorithms and Musical Composition: Potential and Current Abilities

AI is making significant strides in musical composition, showcasing its ability to emulate classical music styles and invent entirely new genres.

Music-generating AI applications like AIVA (Artificial Intelligence Virtual Artist) analyze musical data to identify patterns and structures, enabling them to create compositions that closely resemble those crafted by human musicians.

Creative Experimentation: AI composers can create fresh and original compositions that fuse various styles or create entirely new ones, introducing an important novelty into the creation of music (Porter, 2018).

These tools are becoming essential for musicians, producers, and film scores to help create and orchestrate musical compositions.

AI's Impact on Film and Video Production

In the realm of film and video, artificial intelligence is proving to be a powerful tool that enhances both the creative and technical aspects of filmmaking.

Scriptwriting and Story Development: Smart technologies can analyze existing scripts and provide recommendations about plot, character growth, and dialogue (Dhillon, 2023).

Post-Production Enhancements: AI is used in editing to manage the flow of scenes based on the intended emotions. These tools analyze visual and audio components to dynamically and effectively enhance storytelling.

Visual Effects (VFX) and Color Correction: Since it can take over time-consuming tasks like rotoscoping (tracing over footage frame by frame) and color grading, AI makes the production of visual effects faster and more cost-effective (Davenport & Bean, 2023).

Similar to innovations they are introducing into other sectors, algorithmic applications are used in filmmaking for planning schedules, predicting the public's reaction to a particular movie, and choosing the actors for a cast based on their past performances and audience expectations.

As a result, these technologies are reshaping how we create content, music, and film. They democratize creative processes and enable more people to engage in creative fields, resulting in more diverse art forms.

AI in Gaming

Virtual Reality Meets AI in Games

Artificial intelligence dramatically impacts gaming, especially the branch that integrates virtual reality (VR). It introduces critical advancements that increase engagement and make the experience feel more true to life.

- **Dynamic Interactions:** AI algorithms can animate VR environments in real time with high complexity in response to a player's actions, making the virtual world seem more responsive and similar to reality.
- **Intelligent NPCs:** In VR, AI creates NPCs (Non-Player Characters) with complex and realistic behaviors that can adapt to a player's actions. They use natural dialogues and can learn from a player's strategies (Headleand et al., 2015).
- **Environmental Adaptability:** AI in VR can manipulate aspects of the environment and modify them in reaction to the protagonist's actions or other events in the game. Environmental details it influences include changing the weather, using specific sound effects, and even changing the terrain to fully immerse the player.

These technologies make VR more enjoyable and memorable by providing realistic immersion and emotional experiences that are easier to identify with than earlier gaming experiences.

How AI is Impacting Game Design

The role of intelligent technologies in game design is more crucial than ever, influencing critical aspects of creating and playing games.

- **Procedural Content Generation:** AI creates maps, levels, puzzles, and storylines. Thanks to this, the same game can be different every time it is played, avoiding repetition and keeping players challenged and interested.
- **Adaptive Difficulty:** Gaming challenges can be adjusted according to the skill levels of particular players (Porokh, 2023).

- **Smart Game Balancing:** AI considers players' statistics to maintain the equilibrium of the game mechanics in real time. It can pinpoint aspects that are too challenging or too simple and make necessary modifications to enhance the gameplay experience.

By reshaping the technological and creative aspects of game design, AI can help developers experiment with new approaches to storytelling that can broaden the scope of games as a medium. As they progress with time, intelligent systems are making gaming more interactive, individualized, and immersive.

AI in Society and Administration

Artificial Intelligence is heralding fundamental changes to societal institutions, redefining how people interact and provide services.

AI in Social Interaction

Social Media and Networking

Content Personalization: AI uses big data to learn about user preferences and create popular content for consumers. This increases the amount of time users spend on a platform and introduces them to new areas of interest and communities.

Connection Suggestions: AI can effectively identify similar interests, potential friends, and shared activities to recommend new connections and communities, broadening users' social circles more effectively than traditional search algorithms.

Moderation and Safety: User content is analyzed and moderated with the help of AI, which detects and removes undesirable material. It interprets text and images to filter out content that may be unsafe for communities, ensuring a secure online environment (Srivastava, 2024b).

Technology for the Disabled

Prosthetic and Mobility Aids: AI-enabled smart prosthetics and wheelchairs are self-adjustable and self-controlled, making them more comfortable and effective. Such devices can adapt to a user's movements to enhance function and assistance (Martinez, n.d).

Communication Aids: AI tools that transcribe speech to text and vice versa in real time benefit individuals with speech or hearing impairments. These tools can also help users recognize lesser-known speech patterns over time, thus enhancing their communication skills.

Navigational Assistance: AI technologies can help the visually impaired navigate public spaces by interpreting the environment and providing auditory or tactile signals to avoid hazards (3 Ways AI can help students with disabilities, 2022).

AI continues to improve our social lives by enhancing connectivity, assisting with daily activities, and creating opportunities for disabled people.

Governance, Public Services, and AI

Integrating AI tools into public policy and administration transforms how governments operate and serve their citizens across multiple public sectors.

Public Policy and Administration

Resource Optimization: AI solutions recognize patterns in resource use. This enables more efficient use and the prevention of waste. For instance, AI can help estimate periods of high demand in transport and organize timetables and routes accordingly.

Predictive Governance: Intelligent technologies predict public service requirements and inform governments about potential problems before they arise. Applications range from traffic regulation to public safety measures.

Streamlining Processes: Administrative work such as document processing, license issuance, and recordkeeping uses AI applications to improve efficiency, accelerate procedures, and make them more transparent and less prone to human error (AIM Research, 2023).

AI in the Judicial System

Automated Document Analysis: Large volumes of legal papers can be categorized and reviewed by AI, saving considerable time for lawyers and judges.

Predictive Analysis: Smart technologies analyze past data to determine the possible outcomes of legal precedents and help judges and lawyers make informed decisions.

Case Schedule Management: Automated tools help to organize court proceedings, utilize time effectively, and eliminate protracted legal procedures (Schindler, 2024).

AI in Public Health

Disease Outbreak Prediction: Intelligent agents gather data from various health sources to detect disease outbreaks and recommend early preventative measures, thereby helping to curb the spread of illness.

Patient Data Management: To provide insights for strategic public health decisions, aid in accurate disease diagnosis, and plan effective treatments, AI analyzes vast amounts of healthcare data.

Resource Management: Hospital resources like beds, equipment, and supplies can be managed by AI to ensure that they are optimally allocated to meet patient demands (Olawade et al., 2023).

Using AI as a Tool in Psychology

Mental Health Diagnostics: Intelligent technologies can analyze various parameters of speech, writing, and even facial expressions to help diagnose mental health conditions reliably.

Personalized Treatment: Machine learning algorithms can derive patient history and treatment reactions to prescribe effective, personalized mental health treatments.

Therapeutic Chatbots: Virtual self-help applications that use AI, including specialized chatbots, play an auxiliary role in therapeutic counseling. They are available around the clock to respond instantly to users (Kumari, 2023).

In public services, AI is doing much to enhance efficiency and accessibility, demonstrating its growing relevance in delivering responsive, effective, and inclusive public policies and healthcare. While it offers significant benefits in this area, including increasing the availability of efficient services, it raises concerns about privacy, ethics, and equity. As the technology advances, appropriate regulation of AI will be critical to ensure that its benefits continue to be sustainable.

Key Takeaways

Advancements in artificial intelligence technologies are rapidly transforming the landscape of workplaces across multiple sectors due to the innovations and conveniences they offer.

- **Automating Routine Tasks:** AI can take over routine tasks while improving productivity and minimizing human mistakes. Technologies like RPA (Robotic Process Automation) are instrumental in revolutionizing conventional processes.

- **AI in Project Management:** Its ability to predict the timelines and resources required for project operations makes AI an important tool for project managers.

- **Strategic Decision Support:** AI is instrumental in strategic planning thanks to the forecasts and information that it provides to support informed decisions.

- **The Evolution of Work:** AI is generating new roles and skills that reshape the workplace. We can expect exponential growth of new jobs in AI management and ethical aspects of AI.

- **Learning and Development:** Thanks to advancements in AI, professional training is becoming more customized to employees' needs, enabling essential development and technological aptitude so that individuals can stay relevant and well-suited to a company's requirements.

Integrating AI into daily work and administrative routines is introducing a radical transformation into enterprises' organizational, managerial, and developmental processes. By automating repetitive tasks, AI allows people to focus on more challenging and satisfying work, promoting higher satisfaction levels and greater innovation. Cognitive technology in project management and decision support systems enhances the quality of decision-making. As intelligent systems become increasingly prominent, employees must embrace these to stay current with market trends and maintain their career relevance.

6

QUANTUM AI? HOW QUANTUM COMPUTING CAN REVOLUTIONIZE MACHINE LEARNING

Quantum computing represents a significant advancement over conventional data processing, potentially transforming diverse industries such as cryptography, materials science, and AI. This chapter aims to establish the relationship between quantum computing and AI, starting with the basics of machine learning and quantum mechanics, progressing to quantum computing, and exploring its implications for AI. We will also guide potential entrants into this burgeoning field, outlining what they can expect in the current landscape and the near future of quantum AI.

Understanding the Fundamentals of Machine Learning

When we discussed deep learning in chapter three, we noted that it is a branch of machine learning. ML is the process that enables artificial intelligence to learn from data and experience and make intelligent decisions. This chapter will delve deeper into the fundamental components that form the backbone of ML processes.

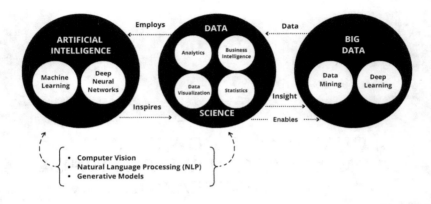

Fig. 8: The Relationship between AI & Big Data

Data Processing

Data used for training must undergo preprocessing before ML models can begin to learn. It is important to understand which steps are integral to this process (Artificial intelligence (AI) vs. machine learning (ML), n.d.).

Data Collection

The first critical step is to gather raw data, which can come from various sources.

- **Databases:** In relational databases, structured data refers to data that is stored and organized according to a predefined structure or model. Relational databases are important for computing and machine learning because they efficiently manage large volumes of structured data, ensuring data integrity and enabling complex queries, which are essential for training and deploying accurate ML models.

- **APIs:** Web services, a type of Application Programming Interface (API), provide real-time information to make a wide range of organizational decisions. They are crucial for data collection and preprocessing in machine learning, as they enable the integration of diverse sources such as weather reports, financial statistics, and social media feeds. For instance, an API can collect real-time stock market information from financial services to train an ML model for predicting stock prices.
- **Sensors:** Data gathered from connected appliances or other tracking systems can be used for ML training.
- **Images and Videos:** Databases and live camera feeds for computer vision applications, such as object detection, facial recognition, and video analysis, can be sources of images and videos.
- **Texts:** Books and other textual resources collected from the internet are used to train natural language processing.

Gathering data from multiple sources ensures that the dataset used for training the model captures most, if not all, of the problem scenarios it is likely to encounter in real-world situations.

Data Cleaning

Before it is used for machine learning, several steps should be taken to improve the quality of a dataset (Artificial intelligence (AI) vs. machine learning (ML), n.d.):

- **Handling Missing Values:** Missing data should be managed by filling in the gaps with average values (mean or median) or removing incomplete data points.
- **Removing Outliers:** Extreme values (outliers) can skew results. As such, they should be identified and removed.

- **Correcting Inconsistencies:** Data must be entered consistently to avoid errors. Large datasets from multiple sources should be systematically organized.
- **Dealing with Duplicate Entries:** Duplicate data must be identified and removed in order to prevent skewed results and maintain accuracy.

These steps are important to generalize the data used in machine learning, ensure that the model does not learn from noise, and prevent it from performing poorly on real-world data.

Normalization/Standardization

Scaling data is essential for some machine learning models, such as neural networks, because these models are sensitive to data magnitudes (Reinaldo & Iano 2021).

Normalization adjusts data values to fall within a specific range, typically between 0 and 1. This process is particularly useful when the data is not evenly spread or follows no specific pattern. By doing this, each data point is scaled down to a consistent range, making it easier for the model to process and learn from it.

Standardization transforms data to have an average value of zero and a consistent spread, or standard deviation, of one. This technique is helpful when the data is more evenly distributed and follows a pattern resembling a bell curve. Standardizing data reduces the impact of extreme values or outliers, ensuring that no single data point overly influences the model's learning process.

Both normalization and standardization help accelerate machine learning and ensure that no single feature or data point dominates the process due to its size. This allows the model to consider all features more equally and learn more effectively.

Feature Selection and Extraction

Selecting and extracting the right features from a dataset is vital to building efficient models (Brown, 2021).

- **Feature selection** refers to choosing which data aspects are most suitable for training the model. Filter methods extract features and use statistical tests for subset ranking. Several approaches can be applied to achieve this. Wrapper methods use other models to rank combinations of features. Embedded methods integrate feature selection into the training process of models like decision trees.
- **Feature extraction** involves converting existing features to reduce the number of dimensions in the data set. This includes creating transformed feature spaces through Principal Component Analysis (PCA) and Linear Discriminant Analysis (LDA).

If these data processing steps are understood and well-executed, the performance of machine learning models is enhanced, making them more efficient and faster.

For example, feature selection might identify the most relevant patient data in a medical diagnosis model, such as age and specific test results. Feature extraction through PCA could reduce the complexity of the dataset by combining related symptoms into single, more informative features. At the same time, LDA could be used to further refine these features based on their ability to differentiate between different diseases, ultimately improving the model's accuracy and speed.

We can look at a counter-example to understand the relevance of PCA and LDA. Suppose feature selection and extraction are not used in a customer recommendation system. In that case, the dataset might include irrelevant or redundant features such as random customer comments or unnecessary purchase history details. This can overwhelm an ML model with noise and irrelevant information, leading to poor performance, slower processing, and less accurate recommendations.

Machine Learning Model Selection

Choosing the right ML model is crucial and depends on the data type and the problem being solved (Brown, 2021).

- **Supervised learning models** work with labeled data (that has known outcomes). They include linear regression, support vector machines, and neural networks. These models predict outcomes (regression) or classify data into categories.
- **Unsupervised learning models** are used when labeled data is not available. Methods like clustering similar data points and association rule mining (finding relationships between variables) help identify hidden patterns in the data.
- **Semi-supervised learning** utilizes a small amount of labeled data combined with a large amount of unlabeled data. This approach is practical when labeling data is expensive or time-consuming.

Once we choose the type of machine learning model we need (supervised, unsupervised, or semi-supervised), the next step is selecting the specific algorithm used for training. Essential factors to consider include the required degree of accuracy, the time it takes to train the model, the size of the dataset, whether the relationship between input and output is linear or nonlinear, and the number of data features (input variables). Each algorithm has strengths and weaknesses depending on a problem's context and limitations.

By understanding these different learning models and the considerations for choosing the most suitable algorithm, we can better address the specific needs of a machine learning project.

Training and Testing

This stage involves splitting the data into training and testing sets, a critical step to prevent overfitting and ensure the model's ability to generalize (Brown, 2021).

Training Phase: During training, the model refines its parameters to minimize errors using methods like gradient descent. This process iteratively adjusts parameters based on prediction errors. For example, gradient descent tweaks parameters like house size or location when estimating house prices to reduce the difference between predicted and actual prices. Effective data preprocessing and parameter optimization result in better-performing models, offering more reliable insights and predictions.

Testing Phase: A trained model is tested with data it has not yet encountered during training. This phase measures the model's performance and ability to generalize on new and unseen data.

Evaluation Metrics

There is an innate tie-in between using evaluation metrics in AI and the connection to quantum computing. Quantum computing can enhance the efficiency and accuracy of machine learning models by processing complex calculations and large datasets faster than classical computers. This can improve the optimization of evaluation metrics, leading to better model performance. Specifically, quantum computing can accelerate the data preprocessing, feature selection, and model training phases that directly impact the evaluation metrics and, consequently, the overall effectiveness of AI applications.

Evaluation metrics are crucial for preprocessing data in machine learning because they provide insights into the model's performance and effectiveness.

- **Classification metrics** such as accuracy, precision, recall (which identifies all relevant cases), and F1 score (which balances precision and recall) assess how well a model sorts data into categories.
- **Regression metrics** like Mean Squared Error (MSE), Mean Absolute Error (MAE), and R-squared, evaluate how well a model predicts continuous values, such as house prices, by accounting for variable noise.

These metrics guide data cleaning, feature selection, and transformation to ensure that AI models learn from relevant information, leading to accurate and reliable predictions in real-world applications. Quantum computing can further enhance these processes by enabling more efficient handling of complex calculations and larger datasets, thereby improving the optimization and effectiveness of these evaluation metrics.

Exploring Machine Learning Algorithms

Artificial neural networks are algorithms that help computers learn from data and make decisions or predictions. Different types of algorithms are used depending on the nature of the data and the specific requirements of an application. While we noted the various kinds of ML algorithms earlier, it is essential to know what defines them.

Overview of Machine Learning

ML Types	Methods	Use Cases
Supervised Learning	Classification Regression	Image Classification, Customer Retention, Diagnostics, Fraud Detection, Forecasting, Predictions, Process Optimization, New Insights
Unsupervised Learning	Clustering Dimensionality Reduction	Structure Discovery, Feature Elicitation, Meaningful Compression, Big Data Visualization, Recommended Systems, Targeted Marketing, Customer Segmentation
Reinforcement Learning	Q-Learning Policy Gradients	Real-time Decisions, Game AI, Robot Navigation, Skill Acquisition, Learning Tasks

Table 2: Types of Machine Learning Algorithms

Supervised Learning

Supervised learning is a fundamental type of machine learning algorithm. In this context, "supervised" refers to a learning process guided by known outputs (labels). Supervised learning aims to develop a predictive model by applying an algorithm to input-output pairs. It operates under the assumption that similar inputs produce similar outputs and uses this relationship to predict outcomes for new, unseen inputs. In what follows, we will look at different examples of supervised learning algorithms (Kumar, 2020).

Linear Regression

Objective: Forecast future values based on one or more predictor variables.

Mechanism: Linear regression estimates the best-fitting line (in two dimensions) or hyperplane (in higher dimensions) that passes through a given set of data points, whereas each dimension represents a feature of the data. Using the least squares method, the optimal fit is achieved by minimizing the sum of the squared differences between the actual data points and the predicted values along the regression line.

Applications: Useful in scenarios where the relationship between predictor variables and the outcome is approximately linear, such as simple trend analysis and basic forecasting tasks.

Logistic Regression

Objective: Predict a binary dependent variable restricted to two distinct categories (e.g., 0/1, true/false).

Mechanism: Despite its name, logistic regression is a classification rather than a regression technique. It estimates the probability of a binary outcome by modeling the relationship between the input variables and the odds of the dependent variable occurring. The dependent variable indicates the likelihood of the outcome being true (or 1), allowing for binary decision-making.

Applications: This method is helpful for email spam detection, disease diagnosis, and other cases where the outcome is categorical.

It helps understand the likelihood of an event or condition based on predictor variables.

Decision Trees

Objective: Use input features, which can be categorical or numerical, to classify something or predict an outcome.

Mechanism: The data is divided into branches at the decision nodes, created at features offering the highest information gain. The branches of the tree represent classification or regression responses.

Applications: Customer segmentation, business rules for operational decision-making, and large datasets where manual categorization is not feasible are among the applications.

Random Forests

Objective: Enhance prediction capability and avoid over-emphasizing one decision tree, which are drawbacks of single decision tree models.

Mechanism: This method grows many decision trees during the training phase and returns the majority class (for a classification task) or average forecast (for a regression task) of the trees. Random forests incorporate randomness into the tree formation, which helps construct a better model.

Applications include gene classification in bioinformatics, credit scoring in finance, and numerous other fields where reliability and accuracy are paramount.

Support Vector Machines (SVM)

Objective: Separate data into two classes by finding the hyperplane that maximizes the distance between them.

Mechanism: SVMs create an n-dimensional hyperplane to separate the data into two classes effectively. They use kernel functions to handle non-linear relationships in the input data, enabling effective classification even when the data is not linearly separable.

Applications: SVMs are used in image classification, text categorization, and other complex environments where the separation between classes is unclear. For instance, they can distinguish between different types of images or categorize text documents based on their content.

Neural Networks

Objective: Used for pattern recognition and to approximate any continuous function.

Mechanism: Neural networks comprise layers of interconnected nodes or digital neurons where every connection denotes a weight. Weights are set using learning algorithms such as backpropagation.

Applications include automated speech recognition, image recognition, and many more instances where AI identifies patterns and data trends in large datasets.

Training and Validation in Supervised Learning

Training involves feeding the model with labeled data, where each input has a corresponding correct output. The model estimates outputs based on the input data and optimizes its parameters by minimizing the error between actual and labeled outputs.

Validation assesses the trained model by using a separate dataset that is distinct from the training data to evaluate its predictive performance and applicability. This process helps optimize model parameters and select the best model from available options.

Unsupervised Learning

Unsupervised learning is a machine learning technique that uses input data without associated output values. It is especially relevant for exploratory data analysis, pattern recognition, and anomaly detection. Let's review the different kinds of unsupervised learning algorithms.

Clustering

Objective: Organize objects so that those belonging to the same cluster are highly similar to one another and dissimilar to those from other clusters to help identify patterns and make sense of large datasets.

Key Methods:

1. **K-Means Clustering** is a process that divides data into a specific number of clusters, assigning each item to the closest cluster mean. It works best for spherical clusters of equal size.
2. **Hierarchical Clustering** builds a tree of clusters without needing to pre-specify the number. It produces a dendrogram, a tree-like diagram that shows how data points are grouped at various levels of similarity to visualize data relationships clearly.
3. **DBSCAN (Density-Based Spatial Clustering of Applications with Noise)** identifies clusters based on data

point density, which is suitable for irregularly shaped data and noisy datasets.

Applications include market segmentation, social network analysis, search result grouping, medical imaging, and image segmentation.

The shapes (spherical, irregular, etc.) in clustering come from how data points are plotted in a multi-dimensional space based on their features. In contrast, each dimension represents a feature of the data. For instance, if you are clustering customer data, one dimension may represent age, another income, etc. Different shapes provide insights into the relationships and patterns within the data, helping choose the right clustering method and interpret results accurately.

For instance, if K-means clustering of customer data with features like age and income shows that the data forms a cylindrical shape, this indicates a strong relationship between age and income. Qualitatively, this suggests that as age increases, income also tends to increase (and then decrease) consistently, showing a clear trend or pattern in the customer group. This could imply that certain age groups have predictable income levels, which can be valuable for targeted marketing and personalized services.

Clustering is crucial for revealing patterns and insights into complex data, aiding decision-making in various fields.

Principal Component Analysis (PCA)

Objective: PCA's purpose is dimensionality (feature) reduction. It reduces the number of variables in a dataset by transforming them into a smaller set of uncorrelated variables while maximizing the variance.

Mechanism: PCA replaces the original variables with new ones formed from linear combinations of the originals, ensuring that the first few new variables (principal components) capture the most significant variations in the data.

Applications include data pre-processing through compression, exploratory data analysis, and noise reduction on finance, biology, and social sciences datasets.

For instance, the first principal component might represent overall academic ability in a dataset with exam scores in math, physics, and chemistry. In contrast, the second principal component could highlight a specific strength in one subject relative to the others. These new variables (principal components) simplify the data, making identifying relevant trends such as general academic performance and subject-specific strengths easier.

Autoencoders

Objective: Train AI to encode datasets to reduce their dimensionality and eliminate noise.

Mechanism: An autoencoder is an unsupervised learning neural network that uses backpropagation. It sets the target values as the input values, aiming to reconstruct its inputs. Data is compressed and then reconstructed.

Applications: By learning efficient data representations, autoencoders facilitate noise reduction and feature extraction tasks. This capability can also support learning in generative models and enable creating new data that mimic the training inputs. Autoencoders can be used for sentiment analysis, detecting anomalies in server usage or website visits, or fraud detection.

A concrete example of using an autoencoder is image denoising. Given a set of grainy photos, an autoencoder can be trained to remove noise. It encodes a noisy image to a lower-dimensional representation and then decodes it to produce a cleaner version, effectively eliminating noise while preserving important features.

Anomaly Detection

Objective: Distinguish data points that do not follow the general trend, known as "outliers".

Techniques:

1. **Statistical Anomaly Detection:** Models the normal behavior of a dataset and uses statistical tests to identify anomalies.
2. **Machine Learning-Based Anomaly Detection** uses clustering or neural networks to learn what constitutes normal behavior and identify deviations.

Applications include fraud detection, network security, fault detection, and system health monitoring.

While these methods can detect anomalies without supervision, autoencoders focus on reconstructing data and finding deviations. In contrast, other techniques like clustering rely on defining and identifying deviations from statistical norms or cluster memberships.

Latent Variable Models

Objective: Extract underlying (latent) structures from data.

Example: Gaussian Mixture Models (GMMs): Gaussian distributions are used to model real-world phenomena that follow their pattern. They represent a normal distribution of variables clustering around a central value with some variability. GMMs estimate parameters of multiple Gaussian distributions that form a mixture. They classify subpopulations within an overall population, even when these subpopulations are not labeled. By analyzing labeled samples and the deviations of other features, GMMs can identify and define unlabeled features, revealing hidden structures in the data.

Applications include voice recognition systems, gene expression profiles, and market research data and analysis.

Evaluation of Unsupervised Learning

Unsupervised learning is more complex to measure than supervised learning since it depends on the specific application and context. Next, we will look at some common methods.

- **Silhouette Scores** measure how similar an object is to its own cluster compared to other clusters.
- **Calinski-Harabasz Indexes** assess clustering quality by comparing the spread (dispersion) between clusters to the spread within them. Higher scores indicate better-defined clusters, meaning that they are distinct and tight.

Unsupervised learning methods are essential for identifying hidden structures in data without prior knowledge of classes or categories. They play a crucial role in data preprocessing, solving complex problems and aiding decision-making across various domains.

For instance, unsupervised learning can group customers based on purchasing behavior without predefined categories in market segmentation, helping businesses tailor marketing strategies more effectively.

Reinforcement Learning

Reinforcement learning (RL) is a type of machine learning where AI agents learn to make decisions in a given environment without receiving explicit instructions. Instead, the agent adjusts its actions based on the consequences (Bernard, 2021).

Core Concepts

Agent: The AI learner or decision-maker.

Environment: The context in which the agent operates, such as a video game the AI is trying to learn how to play.

Action: Possible moves that the agent can make.

State: The current situation as perceived by the agent.

Reward: Feedback from the environment that evaluates the effectiveness of an action. For example, in a video game, gaining points for a successful move or losing points for a mistake.

Policy: The agent's strategy to determine its next action based on the current state.

Value Function: Estimates the expected long-term benefit of a state-action pair.

Model of the Environment: The agent's prediction of how the environment will respond to its actions.

Key Methods in Reinforcement Learning

Q-Learning

Q-learning is a reinforcement learning algorithm used to find the optimal action-selection policy in a finite Markov Decision Process (MDP). An MDP models decision-making with states, actions, rewards, and transition probabilities, where outcomes are partly random and partly controlled by the decision-maker. Q-learning works by learning an action-value function that estimates the expected payoff of taking a particular action in a specific state, and "Q" stands for the quality of the action in terms of its expected reward.

The agent starts with initial Q-values (quantified expected future rewards for actions in given states), which are arbitrary approximations. These values are adjusted as the agent learns from its environment, updating the Q-values based on the rewards received. The goal is to maximize the expected future rewards from each state-action pair until the agent converges to the optimal action-value function. Q-learning is like keeping a record of the best actions tried over time, regardless of whether they were part of the current strategy, to eventually figure out the optimal way to act.

This method is useful when the environment's model is unknown. It is applied in robotics for control and to develop strategies for video games and self-regulating systems requiring sophisticated decision-making without human interference.

Policy Gradient Methods

Policy Gradient Methods are a reinforcement learning algorithm that aims to maximize the policy function by adjusting policy parameters based on the gradient of expected rewards. Unlike techniques focusing on learning value functions, these methods update the policy parameters to maximize the expected reward.

In these methods, the policy is modeled as a parametric probability density that dictates action choices from state observations. The parameters are tuned to achieve the best-expected return, making it suitable for environments with high uncertainty or requiring exploration and allowing for stochastic policies (where actions are chosen based on probabilities). Policy gradients are like continuously improving a single strategy based on recent experiences to maximize rewards directly.

Such methods are ideal for high-dimensional (multi-featured) or continuous policies. They are commonly used in robotic control, autonomous vehicles, strategy games, insurance modeling, natural language processing, and other complex decision-making areas.

Advanced Concepts

Deep Reinforcement Learning combines deep learning with reinforcement learning, using deep neural networks to approximate policies, value functions, or the environment's model. Deep learning works as an integrated component of the RL agent to enhance its ability to process complex, high-dimensional input data and improve its learning and decision-making processes.

Monte Carlo Methods: AI learns directly from episodes of experience without needing prior knowledge of the environment. In the video game example, the AI learns by playing the game multiple times and collecting experiences from each session.

Temporal Difference (TD) Learning: A mix of Monte Carlo and dynamic programming ideas, TD learning gains knowledge of the environment directly from experience. It is suited for real-time updates and ongoing learning processes.

Challenges in Reinforcement Learning

The Credit Assignment Problem refers to determining which actions are responsible for positive or negative outcomes. It is crucial to ensure that the agent learns to associate particular actions with their consequences, enabling it to make better decisions by reinforcing successful behaviors and avoiding unsuccessful ones.

Exploration vs. exploitation is the ability to switch between searching for good actions and repeating those that seemed good in the past. Balancing this is essential for optimizing long-term rewards.

Dimensionality: Handling the complexity of numerous state and action possibilities resulting from many variables (features) is crucial because it directly impacts the reinforcement learning algorithm's ability to explore and learn effectively in complex environments.

These challenges are important because they directly affect the efficiency and effectiveness of reinforcement learning algorithms in real-world applications. Understanding and addressing them is critical to developing robust and scalable RL systems.

Applications of Reinforcement Learning

- **Autonomous Vehicles:** Developing decision-making skills for safe and efficient driving.
- **Finance:** Creating automated trading systems that adapt to market changes.
- **Healthcare:** Providing personalized treatment recommendations based on patient data.
- **Robotics:** Enabling systems to handle objects in dynamic environments.

Reinforcement learning is a type of machine learning used for decision-making. It excels in tasks that require a series of decisions to achieve a specific goal. With advancements in algorithms and computational power, its applications have expanded significantly, solving increasingly complex problems.

Scaling Machine Learning

Scalability in machine learning is important because it enables working with increased quantities of data and more intricate models (Machupalli, 2018).

Big Data Integration

Overview: Tools like Apache Hadoop and Apache Spark help ML algorithms to scale and perform on massive datasets that traditional data processing systems would struggle with, enabling more complex analyses and faster results. They do this by distributing the data processing tasks across multiple machines (Mahendra, 2024).

Techniques:

1. **Apache Hadoop** uses a distributed storage system (HDFS) and a processing framework (MapReduce) that works in parallel to handle big data efficiently.
2. **Apache Spark** enhances this disk-based approach by providing an in-memory engine that significantly speeds up data processing tasks. Spark is suitable when many passes through the data are needed, for instance, in iterative algorithms used in machine learning.

For example, an iterative ML process for filtering emails works like this: A model uses labeled data (spam/not spam) to learn patterns and predict whether new emails are spam. It processes the labeled data repeatedly, calculating errors and adjusting weights based on them. Weights determine the importance of features (e.g. certain words) in predicting spam. This process continues until adjustments no longer significantly reduce the error, resulting in an effective spam filter. The model requires a large amount of training data, and more data improves accuracy, which is why integrating big data is crucially important for ML.

Cloud Computing

In combination with special hardware (cloud processors such as GPUs and TPUs), cloud computing makes training complex models much easier and faster.

GPUs (Graphics Processing Units): Originally designed for rendering graphics, GPUs are highly efficient at parallel processing. They can perform many calculations simultaneously, making them ideal for quickly training complex machine learning models.

TPUs (Tensor Processing Units): Google developed TPUs specifically for machine learning tasks. They are optimized for the high-volume, low-precision calculations used in deep learning and offer even greater efficiency and speed than GPUs for these specific tasks.

Benefits of Cloud Computing

Scalability allows you to easily increase or decrease required computational capacity without the problems of physical hardware requirements.

Accessibility enables ML models to be developed and run from anywhere in the world, making it easier to share and access more resources.

Cost-Effectiveness: Cloud computing reduces the cost of data storage and computation because pay-as-you-go models make computation cheaper.

Both kinds of cloud processors, CPUs and GPUs, leverage massive computational power and can handle multiple operations simultaneously, significantly speeding up the training process for large datasets.

Automated Machine Learning (AutoML)

Overview

AutoML automates data preprocessing to improve the performance of machine learning models, enable more people to use ML, and increase the speed of model deployment.

Capabilities

Data Preprocessing: Implements functions for cleaning and normalizing the collected data.

Feature selection and engineering automatically selects the input variables that are most relevant for predictions and creates new features by identifying patterns in the dataset.

Model selection and hyperparameter tuning utilizes procedures to identify the most suitable machine learning models for a given task and then tune their hyperparameters for optimal performance. Hyperparameters refer to the configuration settings set before the learning process begins, such as the learning rate, the number of trees in a forest, or the depth of a neural network.

Distributed Learning

Overview

Distributed learning divides the ML workload across multiple computing devices using either data parallelism or model parallelism, depending on the problem and system design.

Approaches

Data Parallelism speeds up the training process for large datasets. It divides the data into portions, trains the same model on different portions, and averages the updates.

Model Parallelism is used for very large and complex models. It distributes the model across multiple processors, each handling a segment of the model.

Frameworks: Frameworks like TensorFlow (Google) and PyTorch (Facebook) support distributed learning, making it easy to implement these strategies.

Quantum Computing in AI

Quantum computing represents a next-generation technological advancement, leveraging quantum mechanical phenomena to perform computations unlike classical computing. It can drastically improve the speed, efficiency, and capability of machine learning and AI. By processing large datasets and complex models more quickly, quantum algorithms enhance accuracy and performance in ML. Quantum computing effectively handles highly complex and high-dimensional (multi-featured) data. This section explains how quantum computing complements AI, covering its fundamental principles, potential applications, and real-life expectations.

Quantum Computing Basics

- **Quantum Bits (Qubits):** Unlike conventional computing bits, which can only be in one of two states (0 or 1), qubits can exist in multiple states simultaneously due to quantum superposition, allowing quantum computers to evaluate many possibilities at once (Frankenfield, 2020).
- **Entanglement:** When qubits become entangled, the state of one qubit is directly related to the state of another, regardless of the distance between them. This entanglement enables faster and more complex computations.
- **Quantum Gates:** Similar to logical gates in classical computing, which perform simple logical operations on binary inputs (bits) in order to process data, quantum gates manipulate qubits. However, quantum gates can solve problems that are either impossible or highly impractical for classical computers due to their ability to process multiple states simultaneously.
- **Quantum Decoherence:** A significant challenge in quantum computing is maintaining qubits in their quantum state long enough to perform calculations. Quantum decoherence occurs when qubits lose their quantum properties due to interference from environmental noise.

Quantum AI Applications

- **Quantum Machine Learning:** Quantum algorithms can enhance machine learning by performing computations faster than classical computers, improving efficiency in large and complex models.
- **Optimization problems** can be solved more efficiently by quantum computers than by classical ones, benefiting sectors like logistics, finance, and energy, where data is complex and involves many variables.
- **Drug Discovery:** Quantum computing can analyze and simulate molecular structures beyond the capabilities of classical computers, accelerating the discovery of new medications and therapies.

LEARNING MORE ABOUT QUANTUM AI

Educational Resources

According to one assessment, quantum computing is projected to revolutionize numerous industries, with market value expected to capture nearly $700 billion by 2035 and exceed $90 billion annually by 2040 (McKinsey & Company, 2022). As its applications in AI, optimization, and drug discovery continue to expand, staying informed and educated about quantum computing is essential to leveraging its transformative potential. Various educational resources are available for those keen on exploring quantum AI.

- **Online Courses:** Platforms like Coursera, edX, and MIT OpenCourseWare offer university-level courses on quantum computing basics and their integration into AI systems.
- **Coursera** provides university-level lessons on what quantum computing is and how it is incorporated into AI systems.
- **edX** offers a range of lessons from renowned universities on

quantum computing and its integration into artificial intelligence.

- **MIT OpenCourseWare** provides free MIT courseware, including full lectures on quantum computing and artificial intelligence.
- **Seminars and Workshops:** Periodic seminars and workshops hosted by academic institutions or tech companies provide updates on advances and live demonstrations of quantum computing.
- **Academic Institutions:** Colleges frequently conduct seminars and workshops on current developments in quantum computing.
- **Technological companies:** Many organizations, including IBM and Google, host conferences that involve live demos and new advancements in quantum computing.
- **Periodic Attendance:** Attending such events can help you to become acquainted with new developments and real-world applications.

Textbooks and Online Tutorials

Numerous textbooks and online tutorials present quantum algorithms and their impact on AI, serving as valuable resources for novice and experienced learners.

- Pfaendler, S. M.-L., et al. (2024). *Advancements in quantum computing—Viewpoint: Building adoption and competency in industry.*
- Wong, T. G. (2022). *Introduction to classical and quantum computing.*
- Yanofsky, C., & Mannucci, M. (2008). *Quantum computing for computer scientists.*

Research and Development

As quantum AI continues to evolve, engaging with ongoing research is crucial for everyone, from aspiring computer scientists and AI specialists to business professionals looking to leverage these technologies.

- **Academic Journals:** Read articles in journals like *Quantum Information Processing* and *Quantum Information* to stay informed about new developments and theoretical concepts in quantum AI.
- **Conferences:** Attend conferences such as QIP (Quantum Information Processing) and Quantum AI symposiums, which may be held on-site or online, where researchers present their latest work.
- **Collaborative Research Projects:** Participate in projects offered through university affiliations or cooperative agreements between academic institutions and businesses to gain practical experience and contribute to advancements in the field.

Collaboration

If you are looking to build networks and gain hands-on experience, a number of valuable resources exist.

- **Research Internships:** Renowned tech companies and universities offer internships in quantum computing, providing practical training and professional guidance.
- **Industry Partnerships:** Collaborating with companies developing quantum technologies helps understand quantum AI's commercial and practical aspects.
- **Open Source Projects:** Working on open-source projects like IBM's Qiskit or Google's Cirq can help you explore the real-life applications of quantum computing.

Setting Realistic Expectations

There are high expectations about quantum computing's impact on AI due to its ability to speed up computations and handle larger datasets. The more qubits available, the larger and more complex the datasets AI can process, resulting in higher computational power and potentially greater accuracy. While faster computations can lead to improved energy efficiency for certain tasks, it is important to note that maintaining quantum systems requires significant energy. Future advancements in quantum technology may improve this aspect, making quantum computing more energy-efficient.

UNDERSTANDING THE LANDSCAPE

Acknowledging the current state of the technology

Fault tolerance in quantum computing has yet to be achieved. It is about operating correctly, even in the presence of errors – an essential requirement for reliable and practical quantum computations. Experts estimate that fully fault-tolerant quantum computers may be developed by the 2030s to early 2040s.

Early Stage Technology: It is important to understand that quantum computing is still developing, especially when combined with AI. Realistic, functional solutions to actual problems will require more time to come to fruition. Despite this, there are benefits to using quantum computing in AI now.

Technical Challenges:

- **Probability of Errors:** Quantum computers currently have error rates between 0.5% to 2.9% due to qubit instability and noise (Wang, B., 2024). These errors can affect AI model accuracy and reliability. However, it is possible to mitigate the risks of mistakes made by AI even when using non-fault-

tolerant quantum computing by means of rigorous
validation, testing, and cross-checking.

- **Qubit Coherence Time:** The short coherence time of qubits,
  in which they maintain their quantum state, limits the
  duration for which quantum computations can be reliably
  performed.
- **Engineering Complexity:** Building and maintaining
  quantum computers involves complex engineering
  challenges, including cryogenics and precise control of
  quantum states.

Benefits of Pre-Fault-Tolerant Quantum Computing for AI

Even though quantum computing is still in the early stages of development, it already offers many benefits in combination with AI technologies.

Improved Optimization: Enhances machine learning algorithms
and their training by finding optimal solutions faster, leading to more
efficient AI models.

Accelerated simulations speed up complex simulations used in AI
research, improving model accuracy and development speed.

Hybrid quantum-classical algorithms combine classical and
quantum computing to solve problems more effectively. One prominent example of a hybrid algorithm is VQE (Variational Quantum
Eigensolver), which Google uses in its research to optimize and solve
complex machine learning problems.

Early expertise development provides valuable experience,
preparing researchers for future advancements when fault tolerance
is achieved.

Four prominent examples of how quantum computing is used in
combination with AI include D-Wave's collaboration with Volk-
swagen to optimize city traffic using quantum annealing (a method

for solving optimization problems using quantum mechanics), IBM's Quantum Experience, Qiskit framework for enhancing machine learning algorithms, and Google's Sycamore processor for improving AI tasks like optimization and machine learning.

While quantum computing holds great promise for AI, it is important to set realistic expectations. Technological challenges such as error rates, qubit coherence time, and engineering complexity must be addressed. However, even at this early stage, the benefits of integrating quantum computing with AI underscore the importance of continued innovation and development in this field.

Key Takeaways

In this chapter, we have considered the relationship between quantum computing and the development of AI. Its potential lies in the promise of using quantum computing to significantly increase the speed and efficiency of machine learning, which requires large datasets and substantial computational power to calculate many hypothetical scenarios. To understand this better, we have outlined various applicable ML approaches depending on the nature of the available data and the tasks at hand and will review them now.

Data Processing for Training:

- Collects information from different sources such as databases, APIs, and sensors.
- Cleans data to remove any errors, duplicates, or inconsistencies.
- Preprocesses data to make it ready for machine learning models.
- Chooses significant attributes from the dataset.

Selecting Training Models:

- Supervised learning is employed for categorized data and uses algorithms like linear regression, neural networks, decision trees, and support vector machines (SVM).
- Unsupervised learning is used for unlabelled data. Algorithms include K-means clustering and principal component analysis.
- Semi-supervised learning is utilized when part of the data is labeled.

Reinforcement Learning:

- Understands what an agent, an environment, and an action are.
- Uses techniques such as Q-learning to improve decision-making.

Model Training and Testing:

- Regularizes models to adapt to the data and reduce losses.
- Tests models on new data after training to evaluate their ability to apply their learning to novel information not used during training.

Evaluation Metrics:

- Used to assess a model's performance and ensure that it learns from relevant information during preprocessing in ML.
- Employ classification metrics to provide insights into a model's effectiveness in categorizing data.
- Regression metrics like mean squared error (MSE) and mean absolute error (MAE) are used to correct variable noise and better assess the accuracy of predictions for continuous data.

Scaling Machine Learning:

- Incorporates big data tools such as Hadoop for large datasets.
- Utilizes cloud computing as a flexible resource.
- Employs AutoML for some aspects of machine learning.
- Uses distributed learning to handle large models and data.

Quantum Computing Combined with AI:

- **Enhanced Computational Power:** Quantum computing can significantly speed up complex computations, enabling AI to process larger and more complex datasets more efficiently.
- **Improved Optimization and Training:** Quantum algorithms, leveraging superposition, can explore multiple solutions simultaneously, leading to faster and more accurate optimization and training of AI models.
- **Advanced Simulations:** Quantum computing enables more accurate and faster simulations, benefiting AI applications in drug discovery, climate modeling, and financial forecasting.
- **Scalability for Specific Tasks:** While quantum computing excels in certain optimization and simulation tasks, classical computing is still more scalable and practical for a wide range of general-purpose tasks, including those that require extensive existing software ecosystems and infrastructure. Combining classical and quantum computing allows for developing hybrid algorithms, leveraging both technologies' strengths for better AI solutions.
- **Early Expertise Development:** Integrating quantum computing with AI now provides valuable experience, preparing researchers and developers for future advancements in fault-tolerant quantum systems.

Using quantum computing in combination with AI applications is very promising for the development of the technology. However, quantum computing is still in its infancy and has yet to achieve fault tolerance. Combining this branch of computing effectively with AI will take time and requires further analysis, appropriate goal setting, and continual learning and research.

7

USING AI ETHICALLY AND SAFELY

A s AI technologies develop and more aspects of life are impacted by artificial intelligence, new ethical questions must be addressed and regulated. This chapter explores the major ethical issues concerning AI and how they can be mitigated.

The Ethical Quandary of AI

An Overview of Ethical Problems and AI

Privacy, Surveillance, Manipulation, and Security

The rapid integration of AI into our daily routines raises significant privacy and security concerns that necessitate vigilant oversight and ethical guidelines to protect individual rights and freedoms.

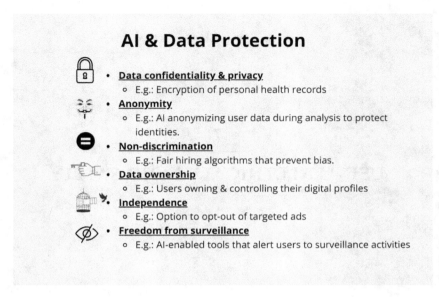

Fig. 9: How AI Can Be Used to Solve Data Privacy Concerns

Privacy Intrusions

Big data and AI have the potential to gather a huge amount of personal data about users. They can create a holistic picture of people's behaviors, preferences, and private lives. Regulation is essential to avoid unethical misuse. Facial recognition and location tracking can invade an individual's personal sphere without consent, raising significant issues about privacy and the dangers of living in a society where our every move and decision is watched (Spair, 2023).

Psychological Manipulation

The use of AI in advertising and social media considerably influences our behavior and decisions. AI's insights into users' habits enable it to recommend content that impacts not only purchasing decisions but also political and social views. This targeted content delivery can exploit psychological vulnerabilities, raising ethical concerns about individual free will, consent, and the potential for manipulation.

Security Vulnerabilities

As with any technology, AI systems are susceptible to cyber security risks, including hacking, phishing, and other cyber threats, which can compromise personal and financial information. This risk can increase when AI is adopted in sensitive sectors, including critical infrastructures and national security systems (Saheb, 2022).

Developing Robust Security Protocols

For these reasons, adequate security measures must be implemented to address the risks of AI technologies. AI systems should use state-of-the-art cybersecurity tools and be designed with security in mind from the ground up. Using a combination of measures like encryption, periodic security assessment, and training AI in secure environments can help organizations protect their systems from breaches.

Regulatory and Ethical Frameworks

To preserve privacy and avoid misuse, it is essential to establish robust regulatory and ethical standards for the responsible use of AI. Governments and international organizations should cooperate in establishing a set of best practices for AI regulation, including topics like data acquisition, user consent, and reporting obligations. These frameworks should reflect the need for AI systems to be developed with privacy and security considerations, including data anonymization and secure data storage.

Bias and Discrimination

As we have seen, one of the most significant ethical challenges associated with artificial intelligence concerns using data for training that is biased or skewed from the outset. AI can learn and perpetuate discriminatory tendencies from it, thereby impacting individuals and groups unfairly.

Forms of Bias in AI

Bias in AI originates mainly from data used in developing these systems. If disparities are present in training data, AI can repeat or potentially exacerbate these biases. We can cite several concrete examples of how this can happen.

If training data do not correctly represent minorities, or if the parameters in the algorithms are not sensitive enough, AI systems used for screening job applicants may lock minorities out of job markets. Another example is facial recognition technology. Research has revealed that AI tends to produce more errors when identifying People of Color than people of European descent, primarily because data used for training often does not reflect sufficiently diverse facial images (Sutaria, 2022).

Eradication of Bias through Technical Approaches

We can employ a variety of technical strategies to mitigate bias in AI systems. To reduce the likelihood of bias, training data must be inclusive. To ensure diverse datasets, as many demographic groups as possible should be represented in the data collection process.

Advanced algorithms must be developed and implemented to detect and address bias in AI decision-making processes. These tools can identify bias by analyzing how an AI operates across different population segments. Procedures that can be affected by biases and identified algorithmically for correction to promote equitable outcomes include processes that determine employee selection, credit approval, and court rulings. Once these biases are detected, another AI system or updated algorithms can be introduced to correct them, ensuring fairness in important social decisions. For instance, IBM's Watson OpenScale identifies biases and suggests corrective actions to guarantee fair AI decisions (Silberg & Manyika, 2019; IBM, 2018).

AI systems should undergo periodic audits to ensure they remain fair and effective. This process includes evaluating outputs for disparities between groups and making necessary adjustments. Fairness can be achieved through manual corrections in specific cases or by retraining the AI with new, more representative data to eliminate systemic bias.

Raising public awareness about potential bias is also crucial for addressing this issue. Educating people about their rights regarding automated decisions empowers them to demand necessary changes and promote fairness in AI systems.

Ethical guidelines and standards should be used to establish values and norms for AI development and deployment. They should encompass data acquisition, algorithm functionality, and oversight mechanisms.

Global Ethical Standards: A Work in Progress

With AI's widespread application, developing global ethical standards is crucial. However, international cultural and legal discrepancies present significant challenges (Golbin & Axente, 2021).

Challenges in Establishing Global AI Ethics

Many factors can lead to international divergence about what the equitable use of AI should mean. The definition of behavior that is considered to be ethical and acceptable can vary between different cultures. For example, surveillance technologies might be more accepted in regions where privacy is less emphasized compared to places where it is highly valued. Attitudes toward ethical guidelines can vary based on the economic or political strategies for AI advancement in different countries. Differences in technological sophistication can lead to varied applications and control of AI technologies from one country to the next. This disparity makes it challenging to set a universally high and achievable standard for all stakeholders.

Efforts Toward International Cooperation

Various international bodies and coalitions have been working toward creating frameworks that can guide the global development of ethical AI. The Global Partnership on AI (GPAI) is an association of professionals from industry, government, and academia dedicated to encouraging the ethical use of artificial intelligence.

GPAI provides recommendations to regulate AI development to support human rights and democratic values. The United Nations has been involved in talks regarding the consequences of artificial intelligence on matters of privacy, security, and ethics to determine how the world should regulate the use of AI. The IEEE Global Initiative on Ethics of Autonomous and Intelligent Systems offers extensive guidelines on the ethical use of artificial intelligence to ensure that it is advanced for the benefit of humanity.

Developing Practical Frameworks

The international discussion about ethical concerns has led to developing practical frameworks and guidelines for artificial intelligence. However, these fall within the realm of "soft law" and represent governance recommendations rather than binding agreements.

These guidelines provide specific recommendations on what the nature of AI should be. They include principles of openness, responsibility, and equity, and offer strategies for implementing these principles during the development and deployment of AI systems.

Some countries have introduced regulatory sandboxes, allowing developers to test AI technologies in a controlled environment, which helps predict the impact of new technologies before they are fully released into the market.

Given the global nature of data, international guidelines on cross-border data transfers, such as the General Data Protection Regulation (GDPR) implemented by the European Union, are crucial. These agreements help build confidence and ensure that AI systems are developed and implemented correctly across borders.

The Role of AI in Shaping Social Norms and Behaviors

Influencing Public Opinion and Behavior

As its implementation and reach grow, artificial intelligence is increasingly influencing social interactions. Specifically, algorithms are deeply integrated into the infrastructure of digital and social media, with significant implications for public opinion and societal behavior. As a technology, AI is neutral. Whether its effect is positive or negative depends on who uses it and with what intentions, making raising awareness about its uses and establishing regulation all the more essential.

Social media websites like YouTube, Facebook, and X (formerly "Twitter") use AI-driven curation and recommendation algorithms to maximize the effectiveness of posted content and user engagement. As such, they tend to support controversial or shocking posts that will attract attention so that users spend more time on the platform. This can mislead the general public by unduly promoting bizarre opinions or fake news, thereby influencing societal attitudes in ways that are not easily identifiable (Shezad, 2023).

Algorithmic optimization of social media content can enhance materials that divide people or intensify their negative emotions, thus contributing to social unrest. For instance, during the Indian elections, misinformation spread through AI-driven social media platforms, including false claims about candidates that fueled political tensions and societal divisions (Mukherjee, 2024; Sebastian, 2024).

Traditional news aggregators use various filters based on user interactions and their predefined interests. This can lead to users being isolated in echo chambers where they are only exposed to views that reinforce their opinions without encountering differing perspectives.

AI can influence user behavior by designing interfaces that guide users towards specific actions. For instance, an online supermarket might encourage behaviors such as making certain purchases. However, it can also be misused to spread misinformation or encourage unhealthy behaviors, thereby having a negative impact. Due to its extensive reach and influence, social media presents particular risks in this regard that users should be wary of.

Experts are calling for education to promote public awareness about the influence of digital media and the need to build resistance to disinformation (Shoaib et al., n.d.). The emotional impact of influential material on social media can exacerbate anxiety and depression, and viewers can fall victim to cyberbullying.

For instance, a study by the University of Vermont found that TikTok's algorithm promotes weight loss and diet culture content, which can negatively impact teens' mental health. The platform's emphasis on weight-normative messaging and repetitive content can lead to disordered eating and unrealistic body image standards among adolescents (Vogel, 2022; Minadeo, 2022).

In dating applications, algorithms control interactions and decide whom people meet based on compatibility results. This may affect personal relationships and enhance stereotyping or bias arising from algorithmic determinations of ideal attributes.

How AI Can Foster an Inclusive Future

Despite any instances of alienation on social media, AI has promising potential to promote inclusiveness. It is especially well suited to create a more inclusive future for learners, students, employees, and consumers.

Personalized Education for Inclusion

We have already seen that AI can revolutionize education by providing personalized learning experiences and adapting to the unique needs of each student. Adaptive, hybrid learning solutions can adjust educational content in real time depending on a student's performance, learning rate, and preferences (Revolutionizing education: The power of adaptive learning platforms, 2023). While any student can benefit from this, supporting individuals with learning deficiencies can be particularly valuable.

Using AI technology, educational content can also be easily translated into multiple languages simultaneously, ensuring that students from different backgrounds can access the same learning material. In addition, AI can play a role in language learning by providing individual lessons and feedback.

Enhancing Accessibility

Intelligent technologies can significantly improve accessibility for people with disabilities—not only by adjusting learning material. They can offer tools that enhance independence and interaction. These include AI voice-to-text converters, smart hearing aids that amplify and filter sounds, and advanced visual recognition systems that provide detailed descriptions of the environment to visually challenged users. Inclusive interfaces that run on AI also benefit users as a visual aid by implementing stark color contrasts. Others provide gesture input modes for users with motor impairments.

Social Inclusion

There are many ways that AI can promote inclusion. As explained, one is auditing for and correcting systemic and dataset biases that can affect the fairness of automated decision-making. Community integration tools powered by AI that can directly promote social inclusion by identifying relevant communities and resources according to user preferences and requirements represent another example. For instance, AI-aided social applications can recommend groups, events, or services within a community to cater to a user's interests or provide access tools to overcome disabilities, thus increasing social inclusion.

Mitigating Risks and Embracing Responsible AI

How Standards Contribute to the Regulation of Ethical AI

We have addressed many reasons why regulation is required for AI. Establishing universally recognized standards is a foundational first step in creating legislation to ensure the ethical development and use of AI technologies.

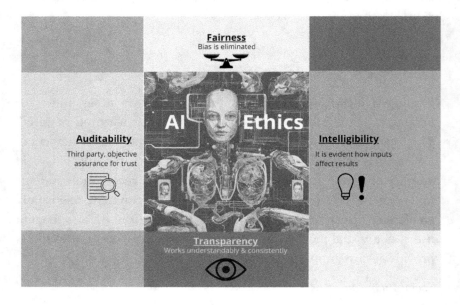

Fig. 10: Essential Elements of Ethics in AI

Clear legal standards governing the application of AI are essential to prevent the unfair treatment of individuals. They should address how data can be processed in AI systems, covering data protection, privacy, and consent. Existing anti-discrimination laws should be extended to AI decision-making, ensuring that individuals can seek redress if an AI system treats them unfairly.

Regulators are mandated to set high standards for AI deployment while ensuring integrity and societal responsibility so that these technologies can safely develop their innovation potential for the benefit of all. Standards can maximize the positive impacts of AI, such as improved healthcare diagnostics and increased efficiency in various industries while minimizing negative consequences like biased decision-making and privacy violations. By establishing rigorous guidelines, regulators can help foster trust in AI systems and encourage ethical development practices.

Preventing Misuse

To deter unethical behavior, standards must clearly define how AI can and cannot be used. They should prohibit activities that could harm society or be used for invasive surveillance, manipulation, or other unethical applications of AI technologies. Defining and enforcing these standards aims to prevent misuse and ensure that AI is developed and deployed in beneficial and safe ways for all.

Setting norms is vital because it creates a basis for legal recourse. Suppose an AI system is deviating from expected performance in a way that potentially risks the safety and wellbeing of individuals or compromises societal values. In that case, well-defined legal processes should be in place to address these issues (Bakiner, 2022).

Guiding Development

Clear standards for using AI guide developers to consider its impact on various aspects of life and how to avoid adverse effects. Ethical guidelines and best practices offer valuable insights and recommendations from experts, helping organizations minimize risks and ensure compliance with the highest standards. This supports the development of AI technologies in ways that are responsible, safe, and aligned with societal values.

A notable example of such efforts is the commitment of eight global tech companies, including Microsoft, LG AI Research, and Lenovo, to adhere to UNESCO's Recommendation on the Ethics of Artificial Intelligence. This framework emphasizes promoting and protecting human rights, diversity, and inclusiveness in AI development and deployment. By integrating these principles, these companies aim to prevent misuse and ensure ethical AI practices (Ramos, n.d.; Eight tech firms vow to build 'more ethical' AI with UN, 2024).

Comprehensive Coverage

Standards for using AI should encompass the entire array of use cases across a range of industries, including healthcare, education, finance, and criminal justice, to ensure that ethical principles are universally applied and not limited to particular sectors. AI technologies will have a global impact. Creating standards for their use enables us to involve different countries and organizations in the regulation process so that no discrepancies arise internationally (Thoms et al., 2023).

Protecting Privacy

Protecting privacy in the age of AI must prevent unauthorized access to users' personal information and ensure that it is handled responsibly. Data protection laws are an essential requirement for these efforts.

Safeguarding Personal Data

With the General Data Protection Regulation (GDPR), the European Union has established a legal best practice to ensure transparency and consent about the use of information. Through it, consumers understand how their personal data is used, and they must agree to it. The GDPR stipulates that only required data be collected and that it be used transparently and only for the purpose for which it was intended. Such provisions can reduce the risk of misuse and even accidental disclosure of sensitive information. Data protection laws should also follow the example of the GDPR in foreseeing that appropriate security measures are established to protect data. These include security audits, encryption, data restoration possibilities, and safe storage solutions.

Security monitoring of private premises or public places is usually done without the express permission of those monitored. While different jurisdictions handle this variously, some cities are following the lead of San Francisco, which became the first city in the US to forbid facial recognition technology for government surveillance (Van Sant & Gonzales, 2019).

In Europe, to maintain compliance with the GDPR, organizations must conduct data protection impact assessments to identify and mitigate potential privacy challenges in specific scenarios early on and safeguard individual rights. These scenarios include using new technologies, large-scale processing of sensitive data, systematic monitoring of public areas, and automated decision-making processes, among others.

European regulation also includes a "right to explanation", allowing individuals to understand the reasoning behind decisions made by automated processing, including AI systems (Selbst & Powles, 2017). The relevant provisions are intended to promote transparency and accountability in the use of personal data to ensure that stakeholders can understand and challenge important decisions made about them, such as those affecting employment or creditworthiness.

Handling Sensitive Information

The GDPR recognizes specific categories of data that require special measures to avoid discrimination and the infringement of individual rights. For instance, these include political opinions and data on health, race, or ethnicity.

According to the GDPR, organizations must report data breaches to the relevant supervisory authority within 72 hours of becoming aware of the violation, unless it is unlikely to result in a risk to individuals' rights and freedoms. If the breach is likely to result in a high risk, the organization must communicate it to the affected individuals without delay to enable them to take protective measures from potential adverse effects (General Data Protection Regulation [GDPR], 2016).

Global Compliance and Cross-Border Data Transfer

Currently, there are no universal global standards for data protection, though several framework initiatives aim to align data protection across borders. The Council of Europe's Convention 108+ seeks to standardize data protection principles internationally. The GDPR has a global impact, applying to any organization processing the personal data of EU citizens, regardless of location. In the USA, local and state laws such as the California Consumer Privacy Act (CCPA) address data protection. Other countries have enacted legislation closely aligned with GDPR, including Canada's PIPEDA (Personal Information Protection and Electronic Documents Act) and Brazil's LGPD (General Personal Data Protection Law). These initiatives are crucial for multinational operations and services involving data processing across various jurisdictions (Beyond GDPR: Data protection around the world, 2021).

International data transfer rules include the Asia-Pacific Cross-Border Privacy Rules (APEC CBPR), Binding Corporate Rules (BCRs) adopted by multinational corporations to ensure compliance with EU data protection standards, and the Standard Contractual Clauses

(SCCs) developed by the European Commission to facilitate data transfers to third countries. GDPR and Convention 108+ also cover the protection of personal data transfers across borders. These frameworks and agreements ensure privacy while facilitating cross-border information flow. Binding regulations like GDPR and BCRs are highly effective, whereas frameworks like APEC CBPR provide voluntary but enforceable standards. Overall, they collectively enhance global data protection despite varying enforcement levels.

Promoting Transparency and Accountability

We have seen that, since AI decision-making is still largely opaque, the technology risks being labeled as a "black box". As its integration into critical industries expands and deepens, achieving transparency about the workings of AI is essential to develop trust in the technology and to ensure that it is used appropriately.

To ensure transparent operations and accountability and build confidence in AI technologies, standards must foresee that users and stakeholders, such as developers, regulators, and impacted communities, can audit and explain decisions made by AI.

Understandable AI

One option to promote transparency is to require all AI systems to be designed with an explanation of the decision-making process in mind. This can be achieved by incorporating visible tools like decision trees into the models with interpretable decision-making parameters.

Developers could be required to provide documentation for audits and regulatory compliance that outlines how an AI system processes data, runs the system, and makes decisions. Third-party monitoring using algorithmic auditing conducted at regular intervals can be made compulsory to review and ensure the fairness, accuracy, and security of AI systems. Audits can reveal biases and issues with the system that are invisible to the people who designed it. Standardizing reporting on how AI operates and achieves its results can increase the understanding of how it works. This could require providing reporting templates about the datasets employed, the models used, and the metrics attained.

Using open standards and engaging in benchmarking can effectively compare and assess different AI systems for fairness, clarity, and efficiency, and stimulate the industry to work on making them more transparent and accountable. User-friendly interfaces that clearly show parameters explaining what the algorithm is doing and the rationale behind particular decisions would also make AI tools more accessible and trustworthy for users.

Contestability

Features that allow users to challenge and dispute decisions made by AI should be incorporated into algorithmic systems. Mechanisms for appeal could include human supervision by line managers enabled to alter the choices made by artificial intelligence if required. Systems should also be designed to incorporate user feedback to improve the consistency and fairness of subsequent results. As we have seen, such feedback loops are beneficial for rectifying mistakes and optimizing the functionality of AI systems.

Encouraging Accountability

Standards regarding AI responsibilities and expected outcomes ensure accountability for its actions, ethical development and use. This is crucial because it allows us to identify who is legally responsible— designers, deployers, or managers—when an AI system causes harm.

This clarity helps in governing the ethical use of AI technologies. Defining legal liabilities to hold organizations responsible for possible damage arising from AI decisions encourages companies to develop effective supervision and accreditation processes for their applications. In parallel to defining and establishing legal responsibilities, AI operators might be asked to have insurance or work with surety bonds through which they can be held responsible for any loss that errors made by AI systems might cause. By requiring AI operators to use such instruments, regulators and customers can mitigate risks associated with mistakes made by artificial intelligence, ensuring accountability and financial recourse in case of harm.

Regulation is an important tool to promote the safe and ethical use of AI technologies. It can address the risks linked to artificial intelligence and support the positive impacts of the technology on society. However, regulation can only succeed in its objectives if backed by strategic planning and instruments that design responsible AI systems aligned with a humanitarian vision and benevolent principles.

Deep Dive: The Ethical Design of AI

STRATEGIES AND POLICIES FOR MITIGATING ETHICAL RISKS IN AI

The first step in managing the ethical risks presented by AI is to develop the technology with foresight for the benefit and wellbeing of all humanity from the start. A number of factors that we have touched on come into play here. Now we will consider the strategies we can use to ensure that they are respected.

Incorporating Human Values

We have established that ethical frameworks must be used as the basis for decision-making by AI in all its processing stages. These must include guidelines for upholding crucial values like privacy, transparency, and fairness. One of the best ways to ensure this is via stakeholder analysis.

When we understand how everyone, including indirect users and those not interacting with an AI system, might be affected, we can better evaluate possible threats and avoid the risk that the technology is misused for personal gain. Such value-sensitive design recognizes the principles involved and considers them from the first stages of conceptual and engineering processes to ensure that the final product upholds the values of the users and society (Cooper, 2023).

Diverse Development Teams

AI systems should be designed, developed, and deployed with input from diverse teams. Such teams are more likely to identify and address biased training data than homogeneous groups. Which specific factors should this diversity include?

Interdisciplinary expertise is extremely important to improve the ability to predict and solve ethical problems. To this end, experts from different disciplines like the social sciences, ethics, law, and the specific industries to which the AI will be applied should be engaged in the development process to bring in different points of view and experiences (Chen, Z., 2023).

Professional diversity is not the only important factor that can help to recognize ethical risks and possible bias in AI. The development team should also represent cultural and demographic diversity, including people of different genders and races from various cultures and social strata.

Implementing Ethical Protocols

Comprehensive reviews are conducted throughout the development lifecycle of an AI before its deployment. They focus on the broader ethical implications of the system, including fairness, transparency, accountability, privacy, and societal impact. They are carried out by stakeholders like ethicists, legal experts, developers, and users, and can include assessments of compliance with ethical guidelines, regulatory requirements, and industry standards. The outcome of an ethical review may result in recommendations for changes, mitigation strategies for potential risks, or even halting a project if ethical concerns cannot be resolved.

Ethical testing examines AI system performance to align it with ethical standards and prevent harmful outcomes. It focuses on behaviors like bias and fairness, robustness against manipulation, and respect for user privacy. Such tests are conducted during development and post-deployment. Techniques include simulation, scenario analysis, and user testing. Outcomes provide data-driven recommendations for improvements, such as algorithm adjustments, retraining models with better data, or implementing new safeguards.

These methods require that developers maintain a transparent documentation log of all the decisions taken during the project, specifically the ethical concerns, methods employed, and reasons for key design decisions. This is important for accountability and may be helpful for audits or meetings with regulatory bodies.

In other words, ethical AI design is not just about doing no wrong but proactively using technology to do good for humanity. By integrating the measures enumerated here, organizations can ensure that AI systems function flawlessly—technically, ethically, and in compliance with cultural norms.

In-Depth Analysis: Audit Types

We have established that AI auditing is important to ensure these technologies consistently perform according to best ethical practices and achieve their intended goals. We should also consider which kinds of auditing methods exist and should be used.

Fairness Audits

Purpose: As we know, counteracting potential algorithmic partiality that may work in favor of or against specific individuals or groups of people is one of the first risks audits must address. These reviews check if decision-making procedures are fair for various groups of people by gender, race, or age.

Methods use feature extraction for statistical and machine learning purposes to detect bias in the outcomes of AI decisions. This may include disparate impact analysis, which aims to identify significant differences in results between groups that should be treated alike.

Outcomes include suggestions for modifying training data or the type of algorithm used, rerunning models, or altering input characteristics to decrease bias.

Accuracy Checks

Purpose: Accuracy tests ensure that an AI system performs correctly and consistently when it encounters new datasets or operates in conditions different than those on which it was initially trained.

Methods include periodically checking the outputs generated by an AI system against a reliable, labeled benchmark or other reference data. This could entail using techniques such as A/B testing in which the current version of the AI system is compared with a previous version to confirm that the latest changes have not compromised the system's efficiency.

Outcomes should maintain the trustworthiness and performance of an AI, which is highly important in fields such as medicine, accounting, and autonomous vehicles, where the impact of algorithmic decisions on people's lives can be decisive.

Security Reviews

Purpose: Security reviews are essential to deter unauthorized persons from acquiring access to or tampering with data used in AI and prevent them from using systems for unwarranted or harmful purposes.

Methods include performing penetration testing to assess the risks in system infrastructure and architecture and engaging ethical hackers to determine the effectiveness of existing security measures. Reviews may also evaluate the physical and electronic security of AI hardware and information management systems.

Outcomes can include additional encryption, better user access controls, and ongoing vigilance for signs of suspicious activity that may indicate a security breach.

Compliance Audits

Purpose: Compliance should be reviewed periodically to ensure that AI systems meet regulatory and legislative requirements, which may differ between countries and sectors.

Methods include checking compliance with applicable regulatory requirements like GDPR for data privacy in the EU or regarding EU citizens, HIPAA for healthcare information in the USA, or other contextually relevant legislation. This comprises documentation review, process validation, and surveys of individuals involved in deploying the AI system.

Outcomes comprise ensuring compliance in AI operations to avoid legal proceedings against deploying organizations and to foster trust between users of the technologies and regulators.

Regularly performing audits helps to achieve compliance with ethical and regulatory requirements and solve possible problems as they arise. It is also a valuable source of insight that can improve the long-term reliability, trustworthiness, and security of AI systems while continuously developing their overall performance.

Risk Assessment

Risk analysis for artificial intelligence is a necessary procedure that enables organizations to understand, evaluate, and control the risks related to these systems. This process involves several specific steps.

Comprehensive Risk Modeling

Objective: Modeling is an essential step in identifying and visualizing all possible threats associated with an AI system during its creation, implementation, and use.

Approaches utilize risk matrices or fault tree analysis to systematically assess and categorize potential risks based on their likelihood and impact. This includes evaluating operational risks for the system, encompassing technical issues such as failures, errors, potential misuse and its social implications.

Outcomes include creating a risk management plan with measures to address threats in the context of the AI system and its particular usage scenarios.

Scenario Planning

Objective: Diverse scenarios should be predicted as part of the modeling exercises to address and respond to possible conditions that might arise and influence the functioning of the AI system.

Approaches entail creating patterns that describe all possible technological advancements, user attitudes, and external factors, including regulatory shifts or market trends. The assessment of each scenario, its likelihood and implications usually includes using both qualitative and quantitative tools.

Outcomes are frameworks that inform the creation of contingency strategies and the establishment of resilient AI systems for various situations.

Privacy Impact Assessments

Objective: To measure the degree to which an AI system safeguards personal data and privacy rights, its impact in these areas should be assessed in advance and on a periodic, ongoing basis.

Approaches entail revisiting data flows in an AI system – including its acquisition, storage, processing, and dissemination – and pinpointing where privacy is at risk of being violated.

Outcomes include data masking, proper data storage, data utilization policies, and other appropriate measures compliant with GDPR or other contextually relevant regulatory requirements.

Stakeholder Engagement

As the implementation and the influence of AI grow, user awareness will become increasingly important to ensure the responsible utilization of these technologies. Engaging stakeholders is integral to the ethical development and deployment of AI, ensuring that the technology aligns with broader social values and the public interest.

Public Consultations

Purpose: To promote democratic participation in the creation and evolution of AI.

Methods: Self-constructed questionnaires, focus groups, and social networks where people share their opinions and fears regarding AI systems.

Impact: Broadens the relevance and applicability of AI systems and improves their acceptance in society by incorporating multiple view-points and requirements.

Expert Panels

Purpose: Domain specialists from various industries should bring in their insights to help define an ethical approach to AI. This is a valuable instrument of public engagement in addition to diverse development teams.

Methods: Weekly or bi-weekly conferences and seminars involving specialists in ethical, legal, technical, and social aspects of AI.

Impact: Expert panels keep AI systems aligned with the most stringent ethical norms and up to date on the latest research and best practices.

Policy Development

Purpose: Developers should work with policymakers to establish and foster rules for applying artificial intelligence and encouraging innovation.

Methods: Involvement in legislative proceedings, offering opinions and evidence to committees and input for policy papers.

Impact: Promotes the development of policies that foster the safety and accountability of AI while promoting its advancement in society.

These complementary risk assessment strategies add up to a practical management framework for AI that addresses technical risks and considers ethical considerations and societal impact.

Continuous Learning and Improvement

One of the most important policies to ensure that artificial intelligence is used ethically is systematically practicing continuous learning. This applies to AI systems, users, and developers alike.

Updating Ethical Standards

Objective: Continuous learning is essential to ensure that developers and AI systems stay abreast of prevailing ethical guidelines, technological advancements, and societal norms.

Approach: Ensure that ethical codes and standards are periodically assessed and updated to reflect the state of current knowledge and tools and align AI systems with them. This can include measures like focus group discussions, expert interviews, or public workshops to capture heterogeneity.

Outcome: Flexible best practices that provide organizations with current principles or rules on how to deal with ethical issues in artificial intelligence.

Lifelong Learning for AI Professionals

Objective: To keep the AI professional up to date and able to implement standards on the use of AI technology.

Approach: Promote lifelong training by providing opportunities for technical and ethical education about AI through workshops, seminars, certifications, and courses.

Outcome: Ensure that developers and IT professionals are both technically skilled and able to address new ethical issues that may emerge.

Adaptive Learning Systems Fueled by Ethical Feedback Mechanisms

Objective: To enable AI systems to evolve, learn, and improve based on their operative experience.

Approach: Introduce features like machine learning feedback loops so that algorithms can continually adjust and improve their ethical alignment when used in the real world. These mechanisms should go beyond existing ones and enable the reporting of ethical issues. Use the insights derived from this feedback to improve the technology.

Outcome: Better AI systems that respond appropriately to ethical issues encountered by users, thereby improving trust and system quality. They ensure compliance with present-day and future ethical norms and can identify and prevent ethical risks.

Integrating Ethical AI Metrics

Objective: To ensure that the ethical components of AI systems are measured and improved over time.

Approach: Develop and incorporate measurement tools that can be used to evaluate the fairness, transparency, and accountability of AI systems. Constantly monitor these metrics to assess progress and identify factors that need improvement.

Outcome: Measurable intel on the ethical behavior of AI systems that can be used to address specific issues and demonstrate the degree of an organization's compliance with ethical norms for the benefit of all stakeholders.

Organizations can use these strategies to effectively manage their AI systems to conform to existing and evolving ethical standards.

Key Takeaways

In spite of its immense potential to improve the quality of human life and advance society, artificial intelligence presents latent ethical risks that must be addressed proactively and continually to ensure the appropriate and benevolent use of the technology.

- **Good governance** is best achieved by instating and continually updating appropriate quality standards, regulatory measures, and policies that enable compliance. Governments, the industry, and organizations that develop the technology are called on here.

- **Global Consensus:** The development of international ethical benchmarks for artificial intelligence is often challenging due to differences in culture and laws. Governments and companies must work together to enable operational consensus.

- **Standards and Guidelines:** Even before globally uniform legislation is in place for AI, standards and best practices, as well as Binding Corporate Rules (BCRs), public commitments, and policies about the ethical development and use of AI technologies can lead the way as guidelines and do much to promote responsible practice.

- **Regional legislation** such as the European Union's GDPR can work beyond borders by requiring compliance from any entity that processes the personal data of EU citizens, regardless of the entity's location. This extraterritorial applicability is one of the central aspects of the GDPR.

- **Awareness of Influence on Public Opinion:** AI can influence social norms and behaviors through content recommendation systems that guide public opinion on social media and elsewhere. Regulatory and policy measures and user awareness about this fact are essential to mitigate risks.

- **Privacy measures** must be established to ensure that AI is not used to invade the personal sphere of users without their consent.

- **Cybersecurity:** AI systems must be designed to address and minimize security challenges from data breaches and misuse for inappropriate surveillance, even when AI is used expressly for monitoring purposes.

- **Bias and discrimination** in AI can contribute to social injustice unless diverse datasets are used from the outset and the quality of training data is continually monitored to remain representative, inclusive, and conducive to fair decision-making. In this connection, auditing, monitoring, and correction can be done by other algorithms.

- **AI decision-making processes** must be kept transparent and explainable, not only for developers but also for third-party audits and users, who should be enabled to contest decisions that they consider unjust by recourse to an objective third party.

- **Transparent documentation** of all AI development phases, mechanisms, and decisions must be made available by developers to auditors and other legitimate stakeholders.

- **Promoting Inclusion:** Just as it is potentially at risk for bias, AI also holds great promise for promoting inclusiveness. For instance, AI-powered educational technologies and supporting devices provide ways to maximize human learning potential and empower diverse users.

Whether the net effect of AI applications is positive or negative is a question of the wisdom and effectiveness of human governance. Addressing ethical issues in AI involves strengthening privacy measures, minimizing biases, and updating the international code of ethics for artificial intelligence. As AI continues to advance, adhering to these ethical principles will be critical to ensure that its development enhances the quality of our lives rather than impeding it.

CONCLUSION

While traversing the domains of generative AI and the many algorithmic use cases, we have traveled across the terrains of machine learning, quantum computing, and their implications in developing advanced AI. We have seen that these technologies are disruptive and can revolutionize industries, improve processes, and foster considerable innovation.

Machine learning is one of the critical components of modern artificial intelligence because it offers effective ways to enhance AI application performance in specific areas. From supervised and unsupervised learning to reinforcement learning, these diverse and versatile methods are central to building intelligent systems capable of efficiently addressing a wide range of tasks.

While it is still an emerging field, quantum computing has the potential to redefine the way artificial intelligence operates. Quantum computing uses the principles of quantum mechanics for information processing. Its integration with artificial intelligence can considerably amplify traditional computation capabilities and pave the way for extraordinarily diverse and valuable applications that can dramatically impact life as we know it.

These AI-powered solutions range from medical diagnostics and remote patient monitoring, which enhance healthcare delivery, to financial forecasting and automated trading systems that can transform the financial industry. AI is revolutionizing transportation with self-driving cars and improving security with facial recognition technologies. It supports interactive and personalized experiences in gaming and movies and enables automation through smart home devices. AI-driven innovations also include bespoke education and AI-enhanced language learning, which tailor learning experiences to individual needs, and automated processes that boost productivity in various industries. Additionally, AI is crucial in research and development, particularly in the pharmaceutical, automotive, and consumer electronics sectors, driving innovation and accelerating product development.

The implementation of generative AI and other types of intelligent systems is heralding a new era with profound impacts and transformative implications across practically all industries. Generative AI enhances the potential of artificial intelligence by creating new content and solutions while enabling applications to learn, make decisions, and collaborate with other intelligent systems and humans. This collective capability maximizes AI's overall effectiveness and adaptability. These advancements can dramatically drive innovation and enhance productivity, significantly improving the quality of our lives and completely reshaping societal structures.

In the near future, these technologies will present us with many more opportunities and challenges. Ethical concerns, the rights to personal data, and the social implications of AI are critical concerns that should not be overlooked or left to chance. Our duty as practitioners and enthusiasts of AI is to meet these challenges responsibly while upholding the principles of the ethical use of artificial intelligence and the commitment to design it for the benefit of society as a whole. In the end, ensuring that this is done right is a task of human governance that should interest all of us.

As we conclude, my goal in this book has been to demonstrate how rapid advancements in AI offer powerful tools to enhance human creativity, efficiency, and well-being. By mastering these technologies responsibly, we can adapt to the digital age and shape it to serve our highest and noblest aspirations. The future holds immense potential, and with thoughtful application of AI, we can ensure it remains a force for positive change and human progress.

CONTINUING THE SPARK OF DISCOVERY

We have reached the end of this exploration into the dynamic world of AI, and I hope I've been able to transform complex ideas into actionable insights. Now that you are well-equipped with the knowledge to innovate and delve deeper, there's one final step you can take to keep this momentum going.

If you've found value in this book, I invite you to share your insights. They could be invaluable to those just beginning their exploration of AI. By leaving a review, you're not only sharing your experience but also guiding others eager to learn. Your thoughts can help them find a clear and approachable introduction to AI, sparking their curiosity and encouraging growth within the AI community.

Simply search for me and this book by title on the platform where you purchased it to leave a review and help others embark on their own AI journey.

Together, we can continue investigating the vast and evolving world of AI, building on the knowledge we've gained and the discoveries still to come.

I look forward to exploring the ever-evolving world of AI with you in my other books!

With gratitude,

Alex Quant

GLOSSARY

Adaptive Learning Paths: Adaptive training content and levels based on the learner's performance and progress through an artificial intelligence application.

Algorithm: A sequence of actions to be followed in order to arrive at a solution to a given problem or accomplish a particular operation.

Algorithmic Transparency: Clarity about how AI systems make decisions.

Anomaly Detection: An Unsupervised Learning Model used to identify outliers or unusual patterns in data that do not conform to expected behavior, often used in fraud detection or monitoring systems.

APIs (Application Programming Interfaces): APIs enable LLM features to be easily incorporated within applications.

Artificial Intelligence (AI): Computer systems designed to simulate human cognitive functions such as learning, reasoning, adaptation, and communication, often using technologies like natural language processing.

Asymptote: A line that a curve approaches but never meets. In calculus, it describes the behavior of functions as the independent variable goes to infinity. In AI, asymptotes help interpret model performance and learning curves. They indicate the limits of improvement and guide optimization by showing where additional data or training may yield diminishing returns.

Attention Mechanisms: Aspects of neural networks that help the model focus on certain parts of the input data to return accurate results.

Augmented Reality (AR): Overlays digital content onto the real world, enhancing the user's perception of their physical environment through devices like AR glasses.

Autoencoders: An Unsupervised Learning Model used for dimensionality reduction and feature learning. It encodes input data into a lower-dimensional representation and then reconstructs the input from this representation.

Automated Graphic Design Tools: AI applications are used to generate and enhance graphic designs.

Autonomous Creation: The ability of LLMs to create textual or visual material independently, such as articles, scripts, or artworks, without human intervention.

Backpropagation: The process of training a neural network by adjusting the weights within the network to reduce error and increase accuracy.

Batch Normalization: A technique used during training to scale inputs to each layer, enhancing stability and training speed.

Bias (in AI): Unintentional skew in a machine learning model caused by biased data samples used for training, which can result in discriminatory or unfair decision-making by the AI.

Big Data: Unstructured data that cannot be processed by conventional data processing software. Big data analytics is essential for determining patterns, trends, and relationships in large datasets.

BERT (Bidirectional Encoder Representations from Transformers): A language model developed by Google that uses bidirectional training to learn context from both sides of a word in a sentence, enabling better text comprehension compared to models with linear dependencies.

ChatGPT: An example of an LLM created by OpenAI that generates human-like text based on input.

Classical Computing: Conventional computation based on Boolean algebra, where bits are binary values (0 or 1). It contrasts with quantum computing, which uses quantum bits or qubits that can represent multiple states simultaneously, allowing for faster and more complex calculations.

Cloud Computing: The use of remote data centers and networks to provision computing resources, such as storage space and processing power.

Clustering: An Unsupervised Learning Model that groups data points into clusters based on their similarities. It has three subtypes: K-Means Clustering, Hierarchical Clustering, and DBSCAN.

Collaborative Features: AI capabilities that enhance teamwork by identifying potential collaborators, assigning tasks, and facilitating interactions.

Computational and Technical Abilities: The ability to analyze and process complex data and create meaningful outputs or insights using computer algorithms.

Computer Vision: A field of AI that enables computers to interpret and understand visual information from the world, such as images and videos, by simulating human vision.

Convolutional Neural Networks (CNNs): Neural networks specialized in handling and discerning visual data, such as images.

Contextual Understanding: The ability of LLMs to establish thematic and stylistic coherence across longer texts by considering the context.

Content Creation AI: AI technologies applied in writing, music, and video creation.

Content Summarization: The process through which LLMs transform long texts into summaries, emphasizing important information and points.

Continuity: A property of a function where it has no jumps, gaps, or holes, meaning its graph is smooth without sharp corners. In mathematics, a function $f(x)$ is continuous at a point $x = a$ if the limit of $f(x)$ as x approaches a equals $f(a)$. Continuity is vital in AI for optimization algorithms. Smooth, continuous functions enable stable and predictable model training, ensuring effective error minimization. Discontinuities can cause erratic behavior in gradient-based methods, impacting model convergence and performance.

Cross-Platform Integration: The compatibility of AI tools with various software applications, enabling integration and efficient operation across different programs.

Data Augmentation: Creating new datasets from existing ones to enhance the training of machine learning algorithms.

Data Confidentiality & Privacy: Ensuring that personal data is securely stored and only accessible by authorized individuals.

Datasets: Collections of text or data provided to AI systems for machine learning. In most cases, datasets for LLMs are large and cover a broad range of topics.

DBSCAN (Density-Based Spatial Clustering of Applications with Noise): An Unsupervised Learning Model that clusters data based on the density of data points, allowing it to identify arbitrarily shaped clusters and handle noise.

Decision Support Systems (DSS): Computerized tools that support business decision-making by analyzing data and providing actionable insights. They come in three types: Data-Driven DSS, which analyzes large datasets; Model-Driven DSS, which uses mathematical models to simulate scenarios; and Knowledge-Driven DSS, which utilizes AI and expert systems to provide specialized advice.

Decision Trees: A Supervised Learning Model that splits data into branches based on feature values, which can be categorical (e.g., yes/no) or numerical. Each branch represents a decision rule, and the process continues until the model arrives at a leaf node that classifies an item or predicts an outcome based on the given data.

Deep Learning: A specific category of machine learning that uses neural networks with multiple layers (deep learning) to solve problems by recognizing patterns in large sets of data.

Deep Fakes: Synthetic media created by AI that resembles real-life videos or audio recordings.

Discriminative AI: AI systems focused on decision-making or classification using data analysis.

Dropout: A strategy to decrease model complexity by temporarily deactivating some neurons during the training process.

Enhanced Visualization: The transformation of data collected through AI into easily understandable graphical forms, enabling executives to gain insights and make informed decisions faster.

Entanglement: A quantum phenomenon in which two or more quantum particles or qubits become linked, and the state of one qubit determines the state of another, regardless of the distance between them.

Ethical Use of AI: Applying AI technologies in a fair, transparent, and responsible manner, with a focus on benefitting society and avoiding harm or exploitation.

Fairness: The ethical consideration of a model's ability to avoid reinforcing or deepening biases that may exist in the training set.

Fine-Tuning: Training an already-trained model on a specific dataset to enhance its performance for a particular task.

Generative Adversarial Networks (GANs): A type of AI model composed of two neural networks, the generator and the discriminator, that engage in a game to produce realistic data.

Generative AI: A subfield of machine learning that produces original content, such as text, images, or music, by training models on large databases.

Generative Capabilities: The ability of LLMs to generate contextually relevant and coherent text based on input prompts.

Generative Pre-Trained Transformers (GPT): A type of deep learning model that uses transformer architecture to generate human-like text based on input prompts. They are pre-trained on large datasets of text and fine-tuned for specific tasks.

Gradient Descent: An algorithm used in machine learning to minimize the cost function by adjusting parameters toward the steepest descent.

Gated Recurrent Units (GRUs): A variation of LSTMs designed to be easier and faster to train, thanks to the incorporation of Forget and Input Gates into a single Update Gate.

Hierarchical Clustering: An Unsupervised Learning Model that builds a tree-like structure of nested clusters by either iteratively merging or splitting clusters based on their similarity.

Human-in-the-loop Systems: AI systems that require human input and supervision to ensure correct and appropriate decisions.

Inclusive and Equitable AI: Ensuring that AI systems are designed and implemented to benefit people of all ages and demographic backgrounds equally.

Interactive Media and Entertainment AI: AI used to produce engaging and personalized media content.

Intellectual Property (IP) Rights: Legal rights that protect creations of the mind, such as inventions, literary works, designs, and artistic works.

Intelligent Algorithms: Computer programs that implement data analysis and artificial intelligence to make decisions and perform tasks.

Intermediate Value Theorem: A theorem in calculus that states that if a function is continuous on the interval [a, b], then there is at least one point ccc in the interval where the function takes that value. This theorem helps ensure smooth optimization landscapes, allowing algorithms like gradient descent to find optimal solutions without abrupt changes or local minima.

Job Redefinition: The change in job descriptions resulting from the implementation of AI, often leading to more analytical and creative tasks for employees.

K-Means Clustering: An Unsupervised Learning Model that partitions data into K clusters by minimizing the variance within each cluster, with each data point assigned to the nearest cluster center.

Knowledge-Driven DSS: Decision support systems that provide specialized problem-solving expertise by analyzing data through AI and expert systems.

Large Language Models (LLMs): Deep learning architectures that both synthesize and interpret natural language from large textual corpora.

Latent Variable Models: A type of Unsupervised Learning Model that infers hidden (latent) variables from observed data, commonly used in factor analysis, topic modeling, and mixture models.

Limit: A key concept in calculus that describes the behavior of a function as its input approaches a particular value.

Linear Regression: A Supervised Learning Model that predicts a continuous outcome by fitting a linear relationship between the dependent variable and one or more independent variables.

Logistic Regression: A Supervised Learning Model used for binary classification tasks, predicting the probability of a categorical outcome, typically represented as 0 or 1, based on input features.

Long Short-Term Memory (LSTM): A type of recurrent neural network (RNN) that can learn long-term dependencies, useful for text generation and prediction tasks.

Machine Learning: A branch of artificial intelligence focused on creating programs that can learn from data and make predictions or decisions.

Multilingual Translation: The ability of LLMs to convey the intended meaning of a text across different languages, enhancing cross-cultural communication.

Multimodal Capabilities: The ability to capture and analyze information from various types of data, including text, images, and audio.

Natural Language Processing (NLP): An area of AI that deals with communication between humans and computers, enabling computers to understand, translate, and generate natural language.

Neural Network: A computational model inspired by the human brain, where data is processed through layers of interconnected nodes (neurons) to identify patterns.

New Skill Requirements: The rise of new technical and soft skills among employees due to the integration of AI technologies in the workplace.

Optimization: In AI, optimization refers to the process of adjusting model parameters to minimize or maximize a specific objective function, such as reducing error or improving performance.

Overfitting: A problem in machine learning where a model becomes too closely fitted to the training data, causing it to perform poorly on new data.

Parameters: Adjustable variables within a model that are fine-tuned to improve accuracy. In LLMs, these parameters can number in the billions.

Pattern Recognition: The capability of AI tools to identify patterns and structures in data that is essential for creating new, realistic content.

Personalization: Adapting content and interactions to enhance the user experience by considering preferences and past behavior.

Policy Gradient Methods: A Reinforcement Learning algorithm that directly optimizes the policy by adjusting its parameters to increase the expected reward, often used in environments with continuous action spaces.

Pre-Training: A broad type of learning where an AI model is trained on a large dataset to gain a general understanding before being fine-tuned for specific tasks.

Predictive Planning: The use of AI to forecast the time required to complete a project and its success rate based on accumulated data and current circumstances.

Predictive Risk Analysis: A risk management approach where AI identifies risks by analyzing data from past projects or similar events.

Principal Component Analysis (PCA): An Unsupervised Learning Model used for dimensionality reduction by transforming data into a set of orthogonal components that capture the maximum variance in the dataset.

Q-Learning: A Reinforcement Learning algorithm that learns the optimal action-selection policy by estimating the value of taking a particular action in a given state, with the goal of maximizing cumulative rewards.

Quantum AI: The application of quantum computing principles to enhance the efficiency and effectiveness of solving complex AI problems, potentially offering superior performance compared to traditional computational methods.

Quantum Computing: A form of computing that uses quantum theory to execute calculations much faster than classical computers for specific problems.

Quantum Gates: Fundamental quantum circuits that operate on qubits, performing operations that underpin quantum algorithms.

Qubit: The basic unit of quantum information, which can exist as both 0 and 1 simultaneously due to superposition.

Random Forests: A Supervised Learning Model that builds multiple decision trees during training and merges them to improve classification or regression accuracy and control overfitting.

Recurrent Neural Networks (RNNs): Neural networks designed to process sequential data, such as time series or language data.

Regulatory Frameworks: Key rules that govern the development and deployment of artificial intelligence technologies to ensure their proper use.

Reinforcement Learning: A branch of machine learning where agents learn to make decisions to maximize cumulative rewards.

Reinforcement Learning from Human Feedback (RLHF): A technique involving human feedback to improve AI model responses and decision-making.

Resource Optimization: The use of AI to allocate resources efficiently, ensuring they are utilized to their full potential.

Robotic Process Automation (RPA): Software configured to mimic and interact with computer systems to perform business processes as if done by a human.

Scenario Simulation: The use of expert systems to model and simulate different business scenarios based on various factors to predict the impact of decisions.

Scalability: The ability of LLMs to process increasing volumes of data while enhancing the capacity and sophistication of the neural network.

Self-Attention: A method in transformer models that adjusts the importance of each word in a sequence, capturing more meaning and improving output.

Squeeze Theorem: A theorem used to determine the limit of a function that is bounded by two other functions with equal limits at a particular point. In AI, this theorem can help in proving the convergence of algorithms and the stability of learning processes.

Statistical Learning: Applying statistical techniques to analyze data and draw conclusions or actions from it.

Strategic Decision Support: Long-term strategic decision-making enhanced by insights and forecasts provided by AI.

Supervised Fine-Tuning: Fine-tuning a pre-trained model using labeled datasets to improve performance on specific tasks.

Supervised Learning: A type of machine learning where the model is trained on labeled data with output labels provided for the training examples.

Support Vector Machines (SVM): A Supervised Learning Model used for classification and regression tasks, which finds the optimal hyperplane that best separates different classes in the feature space.

Task Automation: The use of AI to perform routine and mundane tasks, freeing up human workers to focus on higher-value activities.

Training: The process of feeding data to a machine learning model to allow it to learn patterns and make predictions or create new content.

Transformer Architecture: A neural network architecture that transforms input data using self-attention, enhancing text comprehension and generation.

Transfer Learning: The process of taking a previously trained model and using it for a new but similar task.

Transparency: Ensuring that the decision-making processes of AI models are understandable and explainable.

Unsupervised Learning: A branch of machine learning where the algorithm works on data without labeled responses and attempts to find patterns within the data.

Variance (in ML): A measure of a model's sensitivity to training data, where high variance indicates a risk of overfitting.

Virtual Assistants: Advanced AI systems based on LLMs capable of conversing with users, solving problems, answering questions, and providing customer support. For instance, apple's Siri.

Virtual Reality (VR): An immersive technology that creates a fully digital environment where users can interact with a 3D world using devices like VR headsets.

Visual Data Recognition: AI's ability to identify patterns, objects, and features within images or videos.

XAI (Explainable AI): AI systems designed to provide transparent and understandable explanations for their decisions and actions.

REFERENCES

3 Ways AI can help students with disabilities. (2022, June 3). *Educause Review*. https://er.educause.edu/articles/2022/6/3-ways-ai-can-help-students-with-disabilities

13 ChatGPT limitations that you need to know. (2024, July 12). *Amberstudent*. https://amberstudent.com/blog/post/chatgpt-limitations-that-you-need-to-know

Acerbi, A., & Stubbersfield, J. M. (2023). Large language models show human-like content biases in transmission chain experiments. *Proceedings of the National Academy of Sciences of the United States of America*, 120(44), e2313790120. https://doi.org/10.1073/pnas.2313790120

Addressing bias in AI. (n.d.). *Cte.ku.edu*. https://cte.ku.edu/addressing-bias-ai

Adl, M. (2023, September 11). The role of AI in product development. *Design and Development Today*. https://www.designdevelopmenttoday.com/industries/manufacturing/article/22872677/the-role-of-ai-in-product-development

Ahn, J. C., Connell, A., Simonetto, D. A., Hughes, C., & Shah, V. H. (2021). Application of artificial intelligence for the diagnosis and treatment of liver diseases. *Hepatology*, 73(6), 2546–2563. https://doi.org/10.1002/hep.31603

AI outperforms humans in creativity test. (2023, July 6). *Neuroscience News*. https://neurosciencenews.com/ai-creativity-23585

AIM Research. (2023, July 6). Council Post: The role of artificial intelligence in enhancing public policy. *AIM Research*. https://aimresearch.co/council-posts/the-role-of-artificial-intelligence-in-enhancing-public-policy

Amazon Web Services. (2024). What are large language models? - LLM AI Explained. *AWS Amazon Web Services*. https://aws.amazon.com/what-is/large-language-model/

Artificial intelligence (AI) vs. machine learning (ML). (n.d.). *Microsoft Azure*. https://azure.microsoft.com/en-us/resources/cloud-computing-dictionary/artificial-intelligence-vs-machine-learning

Artificial intelligence (generative) resources. (2024, May 20). *Georgetown University*. https://guides.library.georgetown.edu/c.php?g=1352831&p=9985827

Artificial intelligence: The next frontier in investment management. (n.d.). *Deloitte*. https://www.deloitte.com/global/en/Industries/financial-services/perspectives/ai-next-frontier-in-investment-management.html

Asif, K.J.M. (2023). Interpretable machine learning: Bridging the gap between AI and human understanding. *International Research Journal of Modernization in Engineering Technology and Science*, 5(7). https://www.irjmets.com/uploadedfiles/paper//issue_7_july_2023/43179/final/fin_irjmets1689581659.pdf

Ayuya, C. (2024, June 28). Generative AI vs. AI: Advantages, limitations, ethical considerations. *EWeek*. https://www.eweek.com/artificial-intelligence/generative-ai-vs-ai/

Baheti, P. (2021, July 8). The essential guide to neural network architectures. *V7labs*. https://www.v7labs.com/blog/neural-network-architectures-guide

Bakiner, O. (2022, June 28). Regulation and artificial intelligence ethics: The state of play [Doctoral dissertation, Seattle University]. *Seattle University*. https://www.seat tleu.edu/media/seattle-university/albers-school-of-business/about-albers/center-for-business-ethics/files/Bakiner---Regulation-and-Artificial-Intelligence-Ethics---CBE-White-Paper.pdf

Bartram, S. M., Branke, J., & Motahari, M. (2020, September 10). Artificial intelligence in asset management. *CFA Institute Research Foundation*. https://doi.org/10.2139/ssrn.3510343

Bassel, A., Teixeira, P. E. P., Pacheco-Barrios, K., Rossetti, C. A., & Fregni, F. (2023). Editorial: The use of large language models in science: Opportunities and challenges. *Principles and Practice of Clinical Research, 9*(1), 1–4. https://www.ncbi.nlm.nih.gov/pmc/articles/PMC10485814/

Bernard, C. (2021, December 21). Supervised vs. unsupervised vs. reinforcement learning: What's the difference? *PhData*. https://www.phdata.io/blog/difference-between-supervised-unsupervised-reinforcement-learning

Berryhill, J., Kok Heang, K., Clogher, R. & McBride, K. (2019). Hello, World: Artificial intelligence and its use in the public sector. *OECD Working Papers on Public Governance*. https://doi.org/10.1787/726fd39d-en

Beyond GDPR: Data protection around the world. (2021, May 10). *Thales Group*. https://www.thalesgroup.com/en/markets/digital-identity-and-security/government/magazine/beyond-gdpr-data-protection-around-world

Bird, I. (2023, December 6). Reskilling your workforce in the time of AI. *IBM*. https://www.ibm.com/blog/reskilling-your-workforce-in-the-time-of-ai/

Borges, A. (2023, September 7). Future of art with generative AI. *Ruralhandmade*. https://ruralhandmade.com/blog/future-of-art-with-generative-ai

Brown, S. (2021, April 21). Machine learning, explained. *MIT Sloan School of Management*. https://mitsloan.mit.edu/ideas-made-to-matter/machine-learning-explained

Chacko, A. (2023, October 31). How to use AI writing prompts to get the best out of your AI tools. *Sprout Social*. https://sproutsocial.com/insights/ai-prompt/

ChatGPT for customer service: Limitations, capabilities, and prompts. (2024, January 4). *Klaus*. https://www.klausapp.com/blog/chatgpt-for-customer-service/

Chen, C. (2023, March 9). AI will transform teaching and learning: Let's get it right. *Stanford HAI*. https://hai.stanford.edu/news/ai-will-transform-teaching-and-learning-lets-get-it-right

Chen, Z., Lin, L., Wu, C., Li, C., Xu, R., & Sun, Y. (2021). Artificial intelligence for assisting cancer diagnosis and treatment in the era of precision medicine. *Cancer Communications, 41*(11), 1100–1115. https://doi.org/10.1002/cac2.12215

Chen, Z. (2023). Ethics and discrimination in artificial intelligence-enabled recruitment practices. *Humanities and Social Sciences Communications, 10*(1), 1–12. https://doi.org/10.1057/s41599-023-02079-x

Cheng, R. (2024, February 2). Neuroplasticity in artificial intelligence - unveiling the parallels in AI, the human brain, and beyond. *Linkedin*. https://www.linkedin.com/pulse/neuroplasticity-artifical-intelligence-unveiling-ai-cheng-ruan-md-malsc

Chun, M. (2023, March 20). How artificial intelligence is revolutionizing drug discovery.

Harvard Law Bill of Health. https://blog.petrieflom.law.harvard.edu/2023/03/20/how-artificial-intelligence-is-revolutionizing-drug-discovery/

Conn, R. (2023, July 19). A newbie's guide to generative AI. *Linkedin*. https://www.linkedin.com/pulse/newbies-guide-generative-ai-ryan-conn

Cooper, C. (2023, October 25). How Can We Safely Mitigate the Risks of AI Technologies? *Linkedin*. https://www.linkedin.com/pulse/how-can-we-safely-mitigate-risks-ai-technologies-colin-cooper-ypnff

Cowo, A. (2024, February 26). The influence of AI on social media. *Hive Digital*. https://www.hivedigital.com/2024/02/26/influence-of-ai-on-social-media/

Dane, D. (2023, October 24). 10 Challenges of AI (and how to solve them). *EdApp: The Mobile LMS*. https://www.edapp.com/blog/challenges-of-ai/

Das, S. (2024, February 9). AI tutors: How artificial intelligence is shaping educational support. *ELearning Industry*. https://elearningindustry.com/ai-tutors-how-artificial-intelligence-is-shaping-educational-support

Davenport, T., & Bean, R. (2023, June 19). The impact of generative AI on hollywood and entertainment. *MIT Sloan Management Review*. https://sloanreview.mit.edu/article/the-impact-of-generative-ai-on-hollywood-and-entertainment/

Dempere, J. M., Modugu, K. P., Hesham, A., & Ramasamy, L. K. (2023). The impact of ChatGPT on higher education. *Frontiers in Education*, 8. https://doi.org/10.3389/feduc.2023.1206936

DeSoto, I. (2024). The Art of Algorithms. My Book

Dhillon, S. (2023, July 20). How AI will augment human creativity in film production. *Variety*. https://variety.com/vip/how-artificial-intelligence-will-augment-human-creatives-in-film-and-video-production-1235672659/

Dilmegani, C. (2024a, Jan 2). ChatGPT for customer service: 7 use cases & benefits in 2024. *AI Multiple Research*. https://research.aimultiple.com/chatgpt-for-customer-service

Dilmegani, C. (2024b, June 4). Generative AI copyright concerns & 3 best practices in 2024. *AI Multiple Research*. https://research.aimultiple.com/generative-ai-copyright

Do large language models know what they are talking about? (2023, July 3). *Stack Overflow*. https://stackoverflow.blog/2023/07/03/do-large-language-models-know-what-they-are-talking-about/

Eight tech firms vow to build 'more ethical' AI with UN. (2024, February 5). *TechXplore*. https://techxplore.com/news/2024-02-tech-firms-vow-ethical-ai.html

Elfa, A., Ahmad, M., & Dawood, M. E. T. (2023). Using artificial intelligence for enhancing human creativity. *Journal of Art, Design and Music*, 2(2), Article 3. https://doi.org/10.55554/2785-9649.1017

Everything you need to know about deep learning: The technology that mimics the human brain. (2022, April 21). *Algotive*. https://www.algotive.ai/blog/everything-you-need-to-know-about-deep-learning-the-technology-that-mimics-the-human-brain

Farhud, D. D., & Zokaei, S. (2021). Ethical issues of artificial intelligence in medicine and healthcare. *Iranian Journal of Public Health*, 50(11). https://doi.org/10.18502/ijph.v50i11.7600

Fernández Pérez, I., Prieta, F. d. l., Rodríguez-González, S., Corchado, J. M., & Prieto, J.

(2023). Quantum AI: Achievements and challenges in the interplay of quantum computing and artificial intelligence. In V. Julián, J. Carneiro, R. S. Alonso, P. Chamoso, & P. Novais (Eds.), *Ambient intelligence—Software and applications—13th International symposium on ambient intelligence* (Vol. 603, pp. 177–188). Springer. https://doi.org/10.1007/978-3-031-22356-3_15

Fortino, A. (2023, November 2). Embracing creativity: How AI can enhance the creative process. *NYU School of Professional Studies.* https://www.sps.nyu.edu/homepage/ emerging-technologies-collaborative/blog/2023/embracing-creativity-how-ai-can-enhance-the-creative-process.html

Fraenkel, V., & Kamath, R. (n.d.). In what ways can AI be used for product development? *Delve.* https://www.delve.com/insights/how-collaborating-with-ai-is-trans forming-product-development

Frey, C. B., & Osborne, M. (2023). Generative AI and the future of work: A reappraisal. *The Oxford Martin Working Paper Series on the Future of Work.* https://www.oxford martin.ox.ac.uk/downloads/academic/2023-FoW-Working-Paper-Generative-AI-and-the-Future-of-Work-A-Reappraisal-combined.pdf

The future of creative writing with AI technology. (2024, July 20). *AIContentfy.* https:// aicontentfy.com/en/blog/future-of-creative-writing-with-ai-technology

The future of IT and artificial intelligence. (n.d.). *MyComputerCareer.* https://www. mycomputercareer.edu/news/the-future-of-i-t-and-artificial-intelligence

The future of songwriting: How AI is changing the music composition landscape. (2024, February 13). *On-Page.* https://blog.on-page.ai/ai-songwriter/

General Data Protection Regulation (GDPR), Regulation (EU) 2016/679. (2016). *Official Journal of the European Union,* L119, 1-88. https://eur-lex.europa.eu/eli/reg/2016/679/oj

The generative AI technology stack. (n.d.). *Teradata.* https://www.teradata.com/insights/ ai-and-machine-learning/the-generative-ai-technology-stack

Gerke, S., Minssen, T., & Cohen, G. (2020). Ethical and legal challenges of artificial intelligence-driven healthcare. In *Artificial Intelligence in Healthcare* (pp. 295–336). Science Direct. https://doi.org/10.1016/B978-0-12-818438-7.00012-5

Getting started with prompts for text-based generative AI tools. (2023, August 30). *Harvard University Information Technology.* https://huit.harvard.edu/news/ai-prompts

Golbin, I., & Axente, M. L. (2021). 9 ethical AI principles for organizations to follow. *World Economic Forum.* https://www.weforum.org/agenda/2021/06/ethical-princi ples-for-ai/

Gunter, J. (2023, December 18). How AI content creation is shaping the future of creative writing. *Contentoo.* https://contentoo.com/blog/ai-content-creation-is-shap ing-creative-writing

HarpOnLife: The art of AI prompt crafting: A comprehensive guide for enthusiasts. (2023, November 2). *OpenAI Developer Forum.* https://community.openai.com/t/the-art-of-ai-prompt-crafting-a-comprehensive-guide-for-enthusiasts/495144

Headleand, C. J., Henshall, G., Llyr Ap Cenydd, & Teahan, W. J. (2015). The influence of virtual reality on the perception of artificial intelligence characters in games. *Springer EBooks,* 345–357. https://doi.org/10.1007/978-3-319-25032-8_26

Heller, M. (2019, May 24). What is deep learning? Algorithms that mimic the human brain. *InfoWorld*. https://www.infoworld.com/article/3397142/what-is-deep-learning-algorithms-that-mimic-the-human-brain.html

Hemmler, Y. M., & Ifenthaler, D. (2022). Four perspectives on personalized and adaptive learning environments for workplace learning. In D. Ifenthaler & S. Seufert (Eds.), *Artificial Intelligence Education in the Context of Work. Advances in Analytics for Learning and Teaching* (pp. 27–39). SpringerLink. https://doi.org/10.1007/978-3-031-14489-9_2

The history, timeline, and future of LLMs. (2023, June 26). *Toloka*. https://toloka.ai/blog/history-of-llms/

Holloway, J., Cheng, M., & Dickenson, J. S. (2024, January 13). Will copyright law enable or inhibit generative AI? *World Economic Forum*. https://www.weforum.org/agenda/2024/01/cracking-the-code-generative-ai-and-intellectual-property/

Homolak, J. (2023). Opportunities and risks of ChatGPT in medicine, science, and academic publishing: a modern Promethean dilemma. *Croatian Medical Journal*. 64(1): 1–3. https://www.ncbi.nlm.nih.gov/pmc/articles/PMC10028563/

How can you ensure AI is used responsibly in your team? (n.d.). *Linkedin*. https://www.linkedin.com/advice/0/how-can-you-ensure-ai-used-responsibly-mchce

How can you use AI to automate WFM tasks? (n.d.). *Linkedin*. https://www.linkedin.com/advice/0/how-can-you-use-ai-automate-wfm-tasks-skills-workforce-management-svswf

How can you use AI to personalize employee development? (n.d.). *Linkedin*. https://www.linkedin.com/advice/3/how-can-you-use-ai-personalize-employee-ehvwf

How generative AI is fuelling product development. (n.d.). *Infosys BPM*. https://www.infosysbpm.com/blogs/generative-ai/how-generative-ai-is-fuelling-product-development.html

How is AI impacting IT? (2024, January 22). *Marquette*. https://online.marquette.edu/stem/blog/how-is-ai-impacting-it

How is AI tech like ChatGPT improving digital accessibility? (n.d.). *IA Labs*. https://ialabs.ie/how-is-ai-tech-like-chatgpt-improving-digital-accessibility

How will AI impact information technology: The changing landscape of the IT industry. (2023, June 1). *M&H Consulting*. https://massachusettsitservices.com/blog/how-will-ai-impact-information-technology

Howard, P., & Howard, D. (2023, December 12). IBM Watson OpenScale. *Bloor InBrief IBM*. https://www.ibm.com/downloads/cas/BK0OK0EA

IBM. (2018, February 1). Bias in AI: How we build fair AI systems and less-biased humans. *IBM*. https://www.ibm.com/policy/bias-in-ai/

The impact of artificial intelligence in banking. (n.d.). *ELVTR*. https://elvtr.com/blog/the-impact-of-artificial-intelligence-in-banking

Importance of artificial intelligence (AI) in information technology. (2020, February 18). *Soulpage IT*. https://soulpageit.com/importance-of-artificial-intelligence-ai-in-information-technology

Introduction to deep learning. (2024, May 26). *GeeksforGeeks*. https://www.geeksforgeeks.org/introduction-deep-learning

Jones, E., & Easterday, B. (2022, June 28). Artificial intelligence's environmental costs and promise. *Council on Foreign Relations.* https://www.cfr.org/blog/artificial-intelligences-environmental-costs-and-promise

Kabudi, T., Pappas, I., & Olsen, D. H. (2021). AI-enabled adaptive learning systems: A systematic mapping of the literature. *Computers and Education: Artificial Intelligence,* 2, 100017. https://doi.org/10.1016/j.caeai.2021.100017

Kagan, J. (2024, July 22). How AI can help to increase productivity. *Nifty.* https://niftypm.com/blog/how-ai-can-help-to-increase-productivity

Kayser, D. (2024). 10 best AI project management software you need for 2024. *Forecast App.* https://www.forecast.app/blog/10-best-ai-project-management-software-you-need-for-2024

Klok. (2024, January 22). The impact of AI on the music industry. *Linkedin.* https://www.linkedin.com/pulse/impact-ai-music-industry-klokist-ylqtf

Know it all: ChatGPT and its key capabilities. (n.d.). *SOCO.* https://soco.com.au/know-it-all-chatgpt-and-its-key-capabilities-that-will-surprise-your-team

Kumar, S. (2020, January 29). Supervised vs unsupervised vs reinforcement. *AITUDE.* https://www.aitude.com/supervised-vs-unsupervised-vs-reinforcement

Kumari, J. P. (2023, October 12). AI in psychology - transformative horizons: The integration of AI in psychology services for enhanced mental well-being. *Linkedin.* https://www.linkedin.com/pulse/ai-psychology-transformative-horizons-integration-services-jha-u19vf

Lawton, G. (n.d.). What is generative AI? Everything you need to know. *TechTarget Network.* https://www.techtarget.com/searchenterpriseai/definition/generative-AI

Linegar, M., Kocielnik, R., & Alvarez, M. (2023). Large language models and political science. *Frontiers in Political Science, 5,* Article 1257092. https://doi.org/10.3389/fpos.2023.1257092

Machupalli, V. (2018, March 12). The journey of a machine learning model from building to retraining. *Medium.* https://towardsdatascience.com/the-journey-of-a-machine-learning-model-from-building-to-retraining-fe3a37c32307

Mahendra, K. (2024, February 3). A machine learning project life cycle. *Medium.* https://pub.towardsai.net/navigating-the-exciting-stages-the-journey-of-a-machine-learning-project-life-cycle-db9c187b81c6

Martinez, C. (n.d.). Artificial intelligence and accessibility: Examples of a technology that serves people with disabilities. *Inclusive City Maker.* https://www.inclusivecitymaker.com/artificial-intelligence-accessibility-examples-technology-serves-people-disabilities

Matsuo, Y., LeCun, Y., Sahani, M., Precup, D., Silver, D., Sugiyama, M., Uchibe, E., & Morimoto, J. (2022). Deep learning, reinforcement learning, and world models. *Neural Networks,* 152, 267-275. https://doi.org/10.1016/j.neunet.2022.03.037

McKinsey & Company. (2022, June 25). *How quantum computing could change the world.* https://www.mckinsey.com/featured-insights/themes/how-quantum-computing-could-change-the-world

McKinsey & Company. (2023, August 25). *What's the future of generative AI? An early view in 15 charts.* https://www.mckinsey.com/featured-insights/mckinsey-explainers/whats-the-future-of-generative-ai-an-early-view-in-15-charts

McKinsey & Company. (2024a, March 25). *Driving innovation with generative AI.* https://www.mckinsey.com/capabilities/strategy-and-corporate-finance/our-insights/driving-innovation-with-generative-ai

McKinsey & Company. (2024b, April 2). *What is generative AI?* https://www.mckinsey.com/featured-insights/mckinsey-explainers/what-is-generative-ai

Mearian, L. (2023, February 7). What are LLMs, and how are they used in generative AI? *Computerworld.* https://www.computerworld.com/article/3697649/what-are-large-language-models-and-how-are-they-used-in-generative-ai.html

Merchant, A. (2023, July 29). Bridging the gap between AI and human understanding. *AneesMerchant.* https://www.aneesmerchant.com/personal-musings/explainable-ai-bridging-the-gap-between-ai-and-human-understanding

Michalowski, J. (2022, April 6). An optimized solution for face recognition. *MIT News.* https://news.mit.edu/2022/optimized-solution-face-recognition-0406

Miller, D. (2024, February 26). Guide to optimized AI prompts for internal communications. *Simpplr.* https://www.simpplr.com/blog/2024/optimized-ai-prompts-guide

Minadeo, M., & Pope, L. (2022, November 1). Weight-normative messaging predominates on TikTok—A qualitative content analysis. *PLOS ONE.* https://journals.plos.org/plosone/article?id=10.1371/journal.pone.0267997

Mitchell, M. (2023, July 5). The impact of artificial intelligence on information technology. *MITechNews.* https://mitechnews.com/guest-columns/the-impact-of-artificial-intelligence-on-information-technology/

Montenegro-Rueda, M., Fernández-Cerero, J., Fernández-Batanero, J. M., & López-Meneses, E. (2023). Impact of the implementation of ChatGPT in education: A systematic review. *Computers, 12*(8), 153. https://doi.org/10.3390/computers12080153

Moore, S. (n.d.). Ethical considerations in AI-driven healthcare. *Medical Life Sciences News.* https://www.news-medical.net/health/Ethical-Considerations-in-AI-Driven-Healthcare.aspx

Moret-Bonillo, V. (2015). Can artificial intelligence benefit from quantum computing? *Progress in Artificial Intelligence. 3*(89-105). https://doi.org/10.1007/s13748-014-0059-0

Mukherjee, M. (2024, March 19). AI deepfakes, bad laws – and a big fat Indian election. *Reuters Institute.* https://reutersinstitute.politics.ox.ac.uk/news/ai-deepfakes-bad-laws-and-big-fat-indian-election

Nachaat, M., & Nachaat, M. (2023, February 11). AI Impact on Diagnosis and Treatment. *IEEE Computer Society.* https://www.computer.org/publications/tech-news/community-voices/ai-impact-on-medical-diagnosis-treatment

Najibi, A. (2020, October 24). Racial discrimination in face recognition technology. *Science in the News; Harvard University (SITN).* https://sitn.hms.harvard.edu/flash/2020/racial-discrimination-in-face-recognition-technology

Nazarov, V. (2023, July 10). AI in Telehealth: Revolutionizing healthcare delivery to every patient's home. *TATEEDA.* https://tateeda.com/blog/ai-in-telemedicine-use-cases

Neurovine. (2023, November 10). Building an inclusive AI future: Why diverse voices matter. *Linkedin*. https://www.linkedin.com/pulse/building-inclusive-ai-future-why-diverse-voices-matter-neurovine-xfhzc

Nosta, J. (2023, October 16). Techno-plasticity in the age of artificial intelligence. *Psychology Today*. https://www.psychologytoday.com/us/blog/the-digital-self/202310/techno-plasticity-in-the-age-of-artificial-intelligence

Olawade, D. B., Wada, O. J., Aanuoluwapo, C. D., Kunonga, E., Abaire, O. J., & Ling, J. (2023). Using artificial intelligence to improve public health: A narrative review. *Frontiers in Public Health*, 11(1196397). https://doi.org/10.3389/fpubh.2023.1196397

Ooi, K., Tan, G. W-H., Al-Emran, M., Al-Sharafi, M. A., Căpăţînă, A., Chakraborty, A., Dwivedi, Y. K., Huang, T.-L., Kumar Kar, A., Lee, V., Loh, X.-M., Micu, A., Mikalef, P., Mogaji, E., Pandey, N., Raman, R., Rana, N. P., Sarker, P., Sharma, A., & Teng, C. (2023). The potential of generative artificial intelligence across disciplines: Perspectives and future directions. *Journal of Computer Information Systems*, 1, 1–32. https://doi.org/10.1080/08874417.2023.2261010

Ortiz, S. (2024a, January 23). AI won't replace your job yet because it's still too expensive, according to MIT. *ZDNET*. https://www.zdnet.com/article/ai-wont-replace-your-job-yet-because-its-still-too-expensive-according-to-mit

Ortiz, S. (2024b, June 17). What is ChatGPT and why does it matter? Here's everything you need to know. *ZDNET*. https://www.zdnet.com/article/what-is-chatgpt-and-why-does-it-matter-heres-everything-you-need-to-know

Paul, D., Sanap, G., Shenoy, S., Kalyane, D., Kalia, K., & Tekade, R. K. (2021). Artificial intelligence in drug discovery and development. *Drug Discovery Today*, 26(1), 80-93. https://doi.org/10.1016/j.drudis.2020.10.010

Perwej, Y., & Parwej, F. (2021). A neuroplasticity (brain plasticity) approach to use in artificial neural network. *International Journal of Scientific & Engineering Research*. Article hal-03362902. https://hal.science/hal-03362902/document

Pfaendler, S. M.-L., Kosnon, K., & Greinert, F. (2024). Advancements in quantum computing—Viewpoint: Building adoption and competency in industry. *Datenbank-Spektrum*, 24, 5–20. https://doi.org/10.1007/s13222-024-00467-4

Polding, R. (2023, November 22). How LLMs became the cornerstone of modern AI. *IE Insights*. https://www.ie.edu/insights/articles/how-llms-became-the-cornerstone-of-modern-ai

Porokh, A. (2023, October 19). How will AI disrupt the video game industry in 2024. *Kevuru Games*. https://kevurugames.com/blog/how-ai-is-disrupting-the-video-game-industry

Porter, J. (2018, April 25). The impact of AI on writing and writing instruction. *Digital Rhetoric Collaborative*. https://www.digitalrhetoriccollaborative.org/2018/04/25/ai-on-writing/

Qureshi, R., Irfan, M., Gondal, T. M., Khan, S., Wu, J., Hadi, M. U., Heymach, J., Le, X., Yan, H., & Alama, T. (2023). AI in drug discovery and its clinical relevance. *Heliyon*, 9(7), e17575. https://www.ncbi.nlm.nih.gov/pmc/articles/PMC10302550

Radley, B. (2024, February 20). Using AI to empower an augmented workforce. *Workday Blog*. https://blog.workday.com/en-us/2024/using-ai-empower-augmented-workforce.html

Ramos, G. (n.d.). Ethics of artificial intelligence: The recommendation. *UNESCO.* https://www.unesco.org/en/artificial-intelligence/recommendation-ethics

Reddy, S. (2023). Navigating the AI revolution: The case for precise regulation in health care. *Journal of Medical Internet Research, 25,* e49989. https://www.ncbi.nlm.nih.gov/pmc/articles/PMC10520760/

Reinaldo, P. F., & Iano, Y. (2021). An overview of deep learning in big data, image, and signal processing in the modern digital age. In Y. Iano (Ed.), *Trends in Deep Learning Methodologies.* Science Direct. https://www.sciencedirect.com/topics/computer-science/machine-learning

Revolutionizing education: The power of adaptive learning platforms. (2023, May 24). *Silicon Valley Innovation Center.* https://siliconvalley.center/blog/revolutionizing-education-the-power-of-adaptive-learning-platforms

The role of AI in workforce management processes. (n.d.). *Movo.* https://movo.co/resource/blog/the-role-of-ai-in-workforce-management-processes/

The role of LLMs in AI innovation. (n.d.). *Pecan AI.* https://www.pecan.ai/blog/role-of-llm-ai-innovation

Saheb, T. (2022). Ethically contentious aspects of artificial intelligence surveillance: A social science perspective. *AI and Ethics, 3,* 369-379. https://doi.org/10.1007/s43681-022-00196-y

Samarth, V. (2023, December 14). What is ChatGPT? Top limitations and capabilities you must know. *Emeritus.* https://emeritus.org/in/learn/ai-ml-what-is-chatgpt

Schindler, E. (2024, January 8). Judicial systems are turning to AI to help manage vast quantities of data and expedite case resolution. *IBM Blog.* https://www.ibm.com/blog/judicial-systems-are-turning-to-ai-to-help-manage-its-vast-quantities-of-data-and-expedite-case-resolution

Sebastian, M. (2024, May 16). AI and deepfakes blur reality in India elections. *BBC.* https://www.bbc.com/news/world-asia-india-68918330

Selbst, A. D., & Powles, J. (2017). Meaningful information and the right to explanation. *International Data Privacy Law, 7*(4), 233–242. https://doi.org/10.1093/idpl/ipx022

Shark, A., Shrum, K., Gordon, L., Regan, P., Maschino, K., Shropshire, A. (2019). Artificial intelligence and its impact on public administration. *National Academy of Public Administration.* https://napawash.org/uploads/Academy_Studies/9781733887106.pdf

Shoaib, M. R., Wang, Z., Ahvanooey, M. T., & Zhao, J. (n.d.). Deepfakes, misinformation, and disinformation in the era of frontier AI, generative AI, and large AI models. *School of Computer Science and Engineering Nanyang Technological University Singapore.* https://ar5iv.labs.arxiv.org/html/2311.17394

Shrimankar, D. (2024, February 15). Overcoming challenges in prompt engineering. *Linkedin.* https://www.linkedin.com/pulse/overcoming-challenges-prompt-engineering-shrimankar-pmp-csm-nlkke

Schroer, A. (2024, March 29). Examples of AI in the auto industry: These companies are at the forefront of bringing AI to the automotive industry. *Built In.* https://builtin.com/artificial-intelligence/artificial-intelligence-automotive-industry

Shahzad, A. (2023, November 10). Artificial intelligence and its transformative impact on art, design, and culture: A fusion of innovation and creativity. *Linkedin.* https://

www.linkedin.com/pulse/artificial-intelligence-its-transformative-impact-art-shahzad-ahmad-t4oef

Sharma, S., Rawal, R. S., & Shah, D. J. (2023). Addressing the challenges of AI-based telemedicine: Best practices and lessons learned. *Journal of Education and Health Promotion*, 12(1), p. 338. https://doi.org/10.4103/jehp.jehp_402_23

Shezad, M. (2023, September 5). AI impact on social behaviors. *Linkedin*. https://www.linkedin.com/pulse/ai-impact-social-behaviors-muhammad-shehzad

Shukla, N. (2023, April 27). Evolution of language models: From rules-based models to LLMs. *Appy Pie*. https://www.appypie.com/blog/evolution-of-language-models

Shulman, K. (2024, January 2). The creative future of generative AI. An MIT panel charts how artificial intelligence will impact art and design. *MIT News*. https://news.mit.edu/2024/creative-future-generative-ai-0102

Silberg, J., & Manyika, J. (2019, June 6). Tackling bias in artificial intelligence (and in humans). *McKinsey & Company*. https://www.mckinsey.com/featured-insights/artificial-intelligence/tackling-bias-in-artificial-intelligence-and-in-humans

Singh, A. (2023, July 23). Large language models. *Yellow.ai*. https://yellow.ai/blog/large-language-models

Somers, M. (2023, October 19). How generative AI can boost highly skilled workers' productivity. *MIT Sloan*. https://mitsloan.mit.edu/ideas-made-to-matter/how-generative-ai-can-boost-highly-skilled-workers-productivity

Spair, R. (2023, November 13). The ethics of AI surveillance: Balancing security and privacy. *Linkedin*. https://www.linkedin.com/pulse/ethics-ai-surveillance-balancing-security-privacy-aiethics-spair--sc2pe

Sphinx Solutions Pvt. Ltd. (2024, February 20). Top 5 impacts of AI in banking. *Sphinx Digest*. LinkedIn. https://www.linkedin.com/pulse/top-5-impacts-ai-banking-sphinx-solutions-pvt-ltd--khtef

Srivastava, S. (2024a, July 9). AI in self-driving cars – how autonomous vehicles are changing the industry. *Appinventiv*. https://appinventiv.com/blog/ai-in-self-driving-cars

Srivastava, S. (2024b, July 22). Maximizing engagement: The impact of AI in social media. *Appinventiv*. https://appinventiv.com/blog/ai-in-social-media

Sunday, O. (2023, September 20). Upskilling and retraining in the era of artificial intelligence. *ELearning Industry*. https://elearningindustry.com/upskilling-and-retraining-in-the-era-of-artificial-intelligence

Sutaria, N. (2022, August 29). Bias and ethical concerns in machine learning. *ISACA*. https://www.isaca.org/resources/isaca-journal/issues/2022/volume-4/bias-and-ethical-concerns-in-machine-learning

Sutton, J. (2021, March 8). Artificial Intelligence in Psychology: 9 Examples & Apps. *Positive Psychology*. https://positivepsychology.com/artificial-intelligence-in-psychology/

Sweenor, D. (2023, December 21). Beyond predictions: Exploring the creative and analytical sides of AI. *Linkedin*. https://www.linkedin.com/pulse/generative-ai-vs-traditional-whats-better-david-sweenor-lg16e

Takyar, A. (n.d.). AI in information technology: Use cases, solution and implementation. *LeewayHertz*. https://www.leewayhertz.com/ai-use-cases-in-information-tech nology

Takyar, A. (n.d.). From data to decisions: A guide to the core AI technologies. *Leeway-Hertz*. https://www.leewayhertz.com/key-ai-technologies

Thoms, A., Ehrle, A., & Fischer, K. (2023, November 8). Can a global framework regu-late AI ethics? *Insightplus Baker McKenzie*. https://insightplus.bakermckenzie.com/bm/investigations-compliance-ethics/international-can-a-global-framework-regu late-ai-ethics

Tucker, P. (2024, January 4). How often does ChatGPT push misinformation? *Defense One*. https://www.defenseone.com/technology/2024/01/new-paper-shows-genera tive-ai-its-present-formcan-push-misinformation/393128

University of York. (n.d.). The role of natural language processing in AI. *University of York*. https://online.york.ac.uk/the-role-of-natural-language-processing-in-ai/

Van Sant, S., & Gonzales, R. (2019, May 14). San Francisco approves ban on govern-ment's use of facial recognition technology. *NPR*. https://www.npr.org/2019/05/14/723193785/san-francisco-considers-ban-on-governments-use-of-facial-recognition-technology

Vaswani, A., Shazeer, N., Parmar, N., Uszkoreit, J., Jones, L., Gomez, A., Kaiser, Ł., & Polosukhin, I. (2017). Attention is all you need. *In Proceedings of the 31st International Conference on Neural Information Processing Systems* (pp. 1-11). https://proceedings. neurips.cc/paper/2017/file/3f5ee243547dee91fbd053c1c4a845aa-Paper.pdf

Vogel, K. (2022, November 7). How TikTok perpetuates harmful diet culture among teens, young adults. *Healthline*. https://www.healthline.com/health-news/how-tiktok-perpetuates-harmful-diet-culture-among-teens-young-adults#Lets-recap

Wang, B. (2024, February 7). Quera neutral atom error correction quantum computers. *Next Big Future*. https://www.nextbigfuture.com/2024/02/quera-neutral-atom-error-correction-quantum-computers.html

Wang, C. (2024, January 3). How generative AI is different from traditional AI. *Fivetran*. https://www.fivetran.com/blog/how-generative-ai-different-from-traditional-ai

Wang, H., Zou, J., Mozer, M., Goyal, A., Lamb, A., Zhang, L., Su, W. J., Deng, Z., Xie, M. Q., Brown, H., & Kawaguchi, K. (2024, January 3). Can AI be as creative as humans? *ArXiv.org*. https://doi.org/10.48550/arXiv.2401.01623

Wang, X. (2020). Research on application of artificial intelligence in VR games. In A. J. Tallón-Ballesteros (Ed.), *Fuzzy systems and data mining VI* (pp. 247-254). IOS Press.

Welance. (2023, May 18). ChatGPT and its impact on content creation. *Linkedin*. https:// www.linkedin.com/pulse/chatgpt-its-impact-content-creation-join-welance

Wiggers, K. (2024, January 22). New MIT CSAIL study suggests that AI won't steal as many jobs as expected. *TechCrunch*. https://techcrunch.com/2024/01/22/new-mit-csail-study-suggests-that-ai-wont-steal-as-many-jobs-expected/

Williams, R. (2023, September 14). AI just beat a human test for creativity. What does that even mean? *MIT Technology Review*. https://www.technologyreview.com/2023/09/14/1079465/ai-just-beat-a-human-test-for-creativity-what-does-that-even-mean/

Willmore, J. (2023, December 4). AI education and AI in education. *NSF - National Science Foundation*. https://new.nsf.gov/science-matters/ai-education-ai-education

Wong, T. G. (2022). Introduction to classical and quantum computing. *Rooted Grove.*

Yanofsky, N. S., & Mannucci, M. A. (2008). *Quantum computing for computer scientists.* Cambridge University Press.

Zderic, M. (2023, November 14). 6 key technologies in AI app development. *Decode.agency.* https://decode.agency/article/ai-app-development-key-technologies/

Zhou, E., & Lee, D. (2024). Generative artificial intelligence, human creativity, and art. *PNAS Nexus*, 3(3), pgae052. https://doi.org/10.1093/pnasnexus/pgae052

ABOUT THE AUTHOR

Alex Quant is a PhD, a visionary tech enthusiast who has carved a niche for herself as an authoritative voice in artificial intelligence, quantum computing, futurology and innovations, cryptocurrencies and blockchain. With an extensive background in IT and technology development, she has dedicated her career to exploring the intersections of cutting-edge tech and its implications for the future.

An ardent believer in technology's transformative power, Alex has spent over a decade demystifying complex concepts for the everyday reader. Her passion lies in making complicated technologies accessible and understandable to everyone, from tech novices to seasoned professionals.

Through her writing, Alex is committed to offering insights into the practical applications of rapidly evolving IT innovations. She aims to enlighten her readers about the potential of disruptive tech to revolutionize industries, enhance our personal lives and careers, and address global challenges. Her mission is to empower her readers with the knowledge and tools they need to thrive in an increasingly digital world and recognize the opportunities it offers them.

Beyond her professional work, Alex enjoys early morning walks with her dog, conversing with blue herons, and capturing the intricate beauty of nature through photography.

www.ingramcontent.com/pod-product-compliance
Lightning Source LLC
LaVergne TN
LVHW051444050326
832903LV00030BD/3240